Green India

Requiem to Resurrection

Anand Mohan Rai, IFS (Retd.)
Additional Principal Chief Conservator of Forest
Forest Department
Lucknow, U.P. (India)

International Book Distributing Co.
(Publishing Division)

Published by

INTERNATIONAL BOOK DISTRIBUTING CO.

(Publishing Division)
Chaman Studio Building, 2nd Floor,
Charbagh, Lucknow 226 004 U.P. (INDIA)
Tel. : Off. : 2450004, 2450007, 2459058 Fax : 0522-2458629
E-Mail : ibdco@sancharnet.in

First Edition. 2004

Price: Rs. 595/-

ISBN 81-8189-032-9

Composed & Designed at :
Panacea Computers
33, Nehru Road,
Sadar Cantt. Lucknow-226 002
Phone : 2481164, 2483354, 3127082
E-mail : prasgupt@rediffmail.com

Printed at:
Army Printing Press
33, Nehru Road,
Sadar Cantt. Lucknow-226 002
Phone : 2481164, 2483354, 3246263
E-mail : armypress@indiatimes.com

Preface

A wag narrated a joke about Indian brain. At an auction house, human brains from several countries were kept pre served in jars. When bids were invited, Indian specimen fetched the highest price beating German, Japanese, Russian and US brains hollow. Somebody wanted to know what was so special about Indian brain. The answer was that Indian brain had never been used. The message is loud and clear and applicable for personnel engaged in forestry and environment operations as well. With plenty of rainfall, sunlight, land mass and cheap labour, we have not been able to deliver the goods. The prime reason is failure on the human resource front. Edward Gibbon was once asked about the reason of decline and fall of the Roman Empire after he had finished writing his magnum opus. The short answer was - corruption. Personnel engaged in forestry operations and environmental affairs are so engrossed in lining their own pockets that quality of work suffers badly. New idealistic incumbents soon get disillusioned. They either fall in line with the system or are sidetracked. All this leaves little room for innovative approach and lateral thinking. The stereotyped approach to technology, with shortcuts, result in failed or poor quality plantations with miniscule financial returns. Another drawback is the blinkered vision of persons at the helm of affairs. So-called elites, at the helm of affairs, suffer badly from technophobia. If international donor agencies try to force changes, departments mostly get away by paying lip service. What does Government do when schemes don't work? It simply re-launches them under a new name. Sample this: A particular scheme on women and child welfare has been launched thrice since 2001 under the names — Indira Mahila Yojna, Swayamsidha and Integrated

Women Empowerment Programme. Government schemes, more often than not, turn out to be about mega investments with micro outputs.

In other sectors of economy, the performance of private sector has often been commendable. Unfortunately, some fly-by-night operators exploited the venal instincts of public at large by floating dubious plantation schemes. When the bubble burst, millions of middle class investors were cheated of their hard-earned money. Thus, a golden opportunity was lost to so-called green wastelands - which are, in fact wasted land - by mopping up surplus funds of public at large. The atmosphere has been so vitiated that it will take a lot of entrepreneurship to undertake another such venture and make a success of it. Individual smallholders, however, are quite enthusiastic about planting trees on their land. Even such tree growers are facing problems regarding felling permits and transit permits. Ban on establishing new saw mills is hampering growth of forest-based industries. However, the common man is resigned to overcoming obstacles by palm greasing.

There is threefold ramification for the state of affairs narrated in foregoing paragraphs. Firstly, the nation is deprived of the economic input due from forestry sector in spite of the tract being endowed with abundant natural bounty. This results in weakening of national sinews. *Are they not guilty of playing the role of traitors?* Secondly, we are depriving our present generation and our progeny from benefits that will accrue as a result of environmental amelioration. *Are they not guilty of violating human rights? How different are they from the terrorists we are supposedly fighting.* Thirdly, with the concept of global village, we are reengaging on our international obligations.

Foresters and environmentalists take heart. You are not alone to blame. There are bureaucrats and politicians higher up in the hierarchy. So, the responsibility is collective. At the same time, as the saying goes, it is better to light a candle than to curse the darkness. This book is an effort in that direction. During my association with forest department spanning forty years, one lesson I learnt that square pegs do not fit in the round hole of the system. Also, there is slave mentality within the hierarchy as result of legacy of several centuries. If a worthwhile suggestion comes from within the rank, more likely than not, it is ignored. Same suggestion coming from a foreign consultant is welcomed, though implementation leaves much to be desired and the spirit is lost. As the saying goes, a prophet is seldom respected in his own country. After quitting the government sector, I am feeling like Prometheus* unbound. So the first chapter of this book is "Prometheus Bound " to point out the factors that are inhibiting satisfactory performance in the forestry and environmental sectors. Subsequent chapters deal with simple techniques that can dramatically improve the quality of work regarding greening of India. Some likeminded persons had an afflatus as a result of which we are embarking on a course to implement the ideas propounded in this book through *Badrish Van Mission*. This organization will work as a "social entrepreneur" to unleash a social spark in a businesslike atmosphere to make fundamental shifts in society's established patterns. One of the objectives of this mission will be to set up state of the art **Rural Technology Parks (RTP)** for different agro-climatic zones. These parks will showcase some of the novel ideas put forward in this book and undertake research, training and consultancy functions in the rural sector. Even private sector undertakings will be encouraged to showcase their technologies relevant to the disciplines of rural development. Ten

---

* A character from Greek Mythology

basic fields will find place in the RTPs viz. ethics, dissemination of knowledge, water, land use (including agriculture, horticulture, forestry and medicinal herbs), fuel, value addition in food and other industries, handicrafts, sanitation, ecotourism and rural housing. One of the objectives of the mission will be to popularize locally available wild edible fruit trees. A number of non-traditional fruits are as good a source of nutrients as traditional fruits. These fruits are easier to cultivate and are naturally hardy, requiring less intensive care. Myrica esculenta (Kaiphal) can be used for production of jam, squash and syrup. Fig (Ficus carica), Yellow raspberry (Rubus ellipticus), Wild apricot (Prunus armenica), Malus baccata *(jangli seb),* Diospyros lotus *(amlok)* and Hippophae rhamnoides *(dhurchuk)* are some other wild fruits with considerable medicinal and nutritious value. With the blessings of God Almighty the results will be for all to see on our website www.badrishvan.com in due course.

How to wean away the zombies from their thralldom? Knowledge is power. Gutenberg has rightly been hailed as "The Man of the Millennium". The invention of the letterpress brought in revolutionary changes. Printed products which were luxury items before this invention became mass goods which everyone could afford. A similar state of upheaval and revolutionary changes has currently engulfed the entire media world with the introduction of Internet. "Knowledge" is becoming accessible to all who can handle this tool. Online books can save a lot of paper which is a product of forests. Online books mean different things:

- Printed books that the supplier sells over the internet
- Book files bought online that can be read on the monitor
- Books in digital format that can be deciphered through CD-ROM drive

When multimedia was chosen as the word of the year in 1995, the way forward for books and magazines was mapped out. Electronic reference works clearly offer the consumer more than simple book – speech, sound and animation. Dictionaries and encyclopedias on CD-ROM sell better than software from other sectors. The competition between book and CD-ROM will become fiercer over the next few years. Paper versions of dictionaries and encyclopedias are lagging behind in sale over their electronic half brothers. Internet offers a still better option.

To divagate a little, *Badrish Van Mission* will stay clear of all controversial issues till it finds its moorings so that the Mission is not led astray by detractors.

In general, effort has been to give an overall picture of the scenario with indications about possible future course of action. Wherever needed, essential details have been incorporated to give an insight into off-beat ideas which are not accessible in forestry literature being referred by foresters in India. Otherwise, passing references have been made as the details can be obtained in plethora of manuals and books being churned out. Agricultural matters have been dealt in some detail, reason for which is to be deciphered in following extract from an article by GKB in Indian Forester of September 1989.

> "To illustrate my meaning and intention more plainly, I would compare the relationship existing between Forestry and Agriculture to that which exists in ordinary life between husband and wife. Forestry stands in place of husband, dark, stern and strong, but protecting and cherishing ; Agriculture, bland, benignant and bountiful, may, in my parable be described as the wife. Deprived of the aid and resources derived from the forests, agriculture pines and languishes, and becomes barren and unfruitful"

Lot of water has flown down the Ganga since then. With women's emancipation conjugal relationship has undergone subtle changes. The Israeli experience was not available as a role model. Still, the inference holds good even today for a country like India.

A.M. Rai

1-11-2003

amrai@axsworldmail.com

CONTENTS

Sl. No.	Name	Pages No.
	Preface	
1.	Prometheus bound	1
2.	Water for life	10
3.	Let us turn to that lowly worm	20
4.	Resort to external inputs	25
5.	Poor man's timber can alleviate poverty	33
6.	Native v/s Exotics	41
7.	Breeding Tells	48
8.	Go for multi tier planting	65
9.	Forest fires are manageable	78
10.	Do not be dogmatic about wildlife and vermin	87
11.	Holy cow and other quadrupeds	93
12.	No need to grope in dark	98
13.	Wanted spin doctors- anybody willing to pick up the gauntlets	105

1 Prometheus Bound

From Kautilya to Druker the importance of sense of integrity for those engaged in public affairs has been highlighted. Druker goes so far as to state that a corrupt person is more dangerous for an organization than an inefficient person because a corrupt person corrodes the organization as well. If we scan the history of free India, Nehru in the idealism of youth had declared war on black marketers. On assuming power tainted persons like Krishna Menon and K.D.Malviya were shielded. Indira Gandhi went a step further in defence of corruption calling it a worldwide phenomenon. One PM opined during 1980s that out of every rupee released for public utility, only 85 paisa is utilized for the purpose it is intended. During 1990s another PM revised the figure of 85 paisa to 95 paisa. Forests are the precious endowment of nature to mankind bearing omnifarious resources for sustenance of living beings. Some of the food from forests remain gourmet delight even for sophisticated people in advanced societies. Forests are the repository of a variety of nutritious food gifted by innumerable species in the form of edible tubers, roots, shoots, leaves, flowers, fruits, nuts, seeds, mushrooms, honey etc. These are besides products like timber and firewood. Environmental benefits are additional. Following table estimates the value of an individual tree.

What Is A Tree Worth In Hard Cash

- Generates oxygen worth Rs 5.5 lakhs
- Creates air pollution control worth Rs 10.5 lakhs
- Recycles fertility and soil erosion controls worth Rs 6.4 lakhs
- Creates shelter for birds and animals worth Rs 5.3 lakhs

- Provides flowers, fruits, medicines and other goods worth Rs 4.3 lakh

---

Total value of a tree is Rs 32 lakhs

Late Chaudhary Charan Singh*, while holding the portfolio of Forest Minister of State of Uttar Pradesh, once remarked that forest rangers were a corrupt lot. Some knowledgeable persons enlightened him about the genesis of the problem. In colonial days, extraction of timber from forests was through the agency of contractors. Offices and residences of forest rangers were mostly located deep inside forest areas away from motorable roads. Contractors often had to visit offices of forest rangers in connection with their contracts. More often than not, there were not even tea stalls in neighboring areas, what to talk of restaurants or hotels. For the sake of social courtesy, forest rangers used to offer tea, snacks, lunch or dinner as per the occasion. In return, contractors started paying small amounts as *nazrana* or gratuity to forest rangers. It was termed as lot fees. There was no element of concept of bribe either for the receiver or the giver. Gradually, there was a change in the state of affairs. Forest areas became more accessible. Means of transportation became more plentiful and faster. Small establishments dispensing eatables started mushrooming. The original **raison** d' **être** for lot fees was no longer valid. But the custom persisted as perk of office for forest rangers. Contractors started expecting official favors in return. Colonial masters turned a blind eye to the state of affairs so long as there were no major transgressions of rules. After independence, matters took a turn for the worse. Rates per lot were customarily uniform. So, more the number of lots in a range, more lucrative it was considered. Power of posting in ranges was with higher ups and they also started expecting a share in the pie. The system started degenerating to such an extent that sometimes

---

* Prime minister of India for some time

bids were invited informally for most lucrative ranges. Once the top bosses were exposed as corrupt, office staff also jumped into the fray.

Some enterprising rangers devised ways and means of increasing number of lots and amount of lot fees - not that such black sheep were rare during colonial rule. For Sal (Shorea robusta) borer insect infested trees the treatment prescribed was – and still is – to fell the affected trees so that healthy trees do not get affected. During 1930s, one enterprising forest ranger sent a report to higher ups about infestation of Sal borer insects. His diligent British boss went to inspect the area on a fact finding mission. The ranger got bore holes drilled in a large number of trees with augers. The resulting sawdust was sprinkled on the ground. The British forester was convinced about large scale borer infestation and issued order for marking of all infested trees for felling. Hundreds of lots were formed of perfectly healthy trees and were duly felled by purchaser contractors. The forest ranger got his cut as lot fees. Later on, after independence, whenever somebody wanted more number of lots than silviculturally admissible, dead, dying, so called diseased trees were marked for felling in the name of maintaining the hygiene of the forest. Apparently, forest officers were oblivious of the role of such trees in overall ecological scheme of things. Persons in highest echelons were also not averse to dramatically enhance the quantum of timber to be felled. All sorts of logical arguments were advanced to clear fell the miscellaneous forests during 1960s to be replaced by fast growing exotics. This benefited forest officials in two ways. Firstly, there were more number of felling lots. Secondly, there was huge flow of funds for planting clear felled areas resulting in one more malaise creeping in – viz. that of preparing false muster rolls and inflated or false expenditure vouchers. Easy money sapped the professional caliber as well. Orientation of forest officials changed from sylvan to urban.

By and by, isolated incidences of illicit felling of trees by contractors – i.e. felling of trees not marked and not included in sale deed – started surfacing. This forced a policy decision to do away with contractors and introduction of government-owned forest corporations. These corporations were, by and large, not commercially oriented with all the trappings of bureaucratic set up. Lot fees dried up with the demise of contractor system but other avenues were devised for lining up the pockets of forest officials. A new term forest mafia was coined to cover up cases of illicit felling. A closer analysis will reveal that more often than not, forest officials indulging in illicit felling needed some front men. Later on, these front men became more powerful than their masters and started operating independently with all the clout that money and muscle power wields in India. With regulations being enforced with a view to enforce greater conservation efforts, prohibited acts are being clandestinely done. For instance, mining of sand boulder etc. are officially prohibited in sanctuaries. But forest officials often run their private quarries. No new sawmill can run beyond those reported to Hon'ble Supreme Court. But some forest and police officials are making a fast buck by just turning a blind eye.

Another ingenious method of fleecing farmers is the Tree Protection Act. Though, as an eyewash, number of exempted species have been increased, still, enough impediments remain for forest officials to extract their pound of flesh. Another lucrative source is Transit Rules. Even exempted species require a transit permit from forest officials in most cases. Forest barriers were devised for checking of illicit forest produce. These barriers have emerged as the most lucrative source of easy money. There are serious consequences of bribe giving and bribe taking. The giver wants to avoid the hassles involved in standing up to principles. By and by, his moral sinews are weakened and when temptation comes in the way, there is little hesitation in committing other heinous crimes. On the other hand, the bribe taker has the wherewithal to undermine the system. Net result is system collapse.

One obvious question that comes to mind is, what about the system of checks and balances. The simple answer is the pipeline of sleaze money runs from bottom to top with attendant protection racket. If somebody dares to choke the pipeline, he pays the price or he is sidelined. Principle of Ricardo is operating – bad money drives the good money out of circulation. Substitute personnel for money and you get the picture. If the slogan of greening of India is to be translated into reality, the system needs overhauling besides technological innovations. After all, manpower is the most valuable asset of any organisation and nation. The endeavour should be to bring the system on right track in our interest, for the sake of our progeny and for humanity. If Prometheus is unbound someday, somehow from the shackles of the unjust and corrupt system, things will improve miraculously.

Knowledge is a powerful tool for liberating the soul of individuals. However, it is a double-edged weapon. In the hands of evil persons, it becomes a tool for oppression. In order to fight evil, good persons have to band together and utilize knowledge as a weapon. Fortunately, there is a chink in the armour of evil persons. They also have to pretend that righteousness is a virtue to be cultivated. The task is to free the zombies from the clutches of oppressors and to force the fence sitters to commit one way or other so that the blurred margin between black and white becomes sharply defined. There are no grey areas blurring the divide. Print media played an important role in ushering social justice during last two centuries. Radio played a similar role during last seven decades. Now we have CDs, DVDs (along with the players and screens), digital video cameras, cable TV, direct to home TV channels, internet, satellite radio broadcasts, digital telephonic systems with land lines as well as mobiles and what not. As per the vision of late Dhirubhai Ambani, majority of Indians cannot decipher written communication even if they are in the category of literates. But, all can follow verbal communication. To go one more step ahead, audio-visual method is the best mode of communication. In order to reach the hearts and minds of target audience, educational CDs, DVDs and telecasts are the best option. So, the first task is to create interesting and appealing teaching software. Experts believe that knowledge management is going to be a major business enterprise in coming decades. Also, the sum total of knowledge is doubling every two years. So, knowledge storage and dissemination also need attention.

Forest officials in particular, and Indians in general will have to shed their technophobia. India is a vast nation with mega diversity. So, any generalization will not be appropriate. Bihar is altogether different from Chadrababu Naidu's Andhra. Again, the urban elites of Andhra are poles apart from the Naxalite affected tribals. Still, if the top executives are willing to adopt cutting edge futuristic technologies, then underlings will have to follow suit. In military science the expression C^3I is used for success in the battlefield. The acronym stands for command, control-communication and intelligence. All these are equally essential for forestry, specially, when the lower-level functionaries are located at remote places. Some efforts were made in some states for quick communication by establishing wireless system. This has not yielded desired results as mostly the systems are out of order. More attention needs to be paid to websites, internet, e-mail, intranet, SMS, video conferencing and voice mail instead of wallowing in the muck of files. The time saved by middle-level functionaries by avoiding personal attendance with bundles of dusty dog-eared files can be better utilized in field inspections, training and learning from farmers and grass-root level functionaries and enhancing knowledge base.

Lot of precious land has been alienated due to forest boundaries being fuzzy. This can easily be avoided by using GIS and satellite imagery. Once the boundary is established on GIS, durable demarcation pillars can be affixed. For so many decades, forest departments have been lamenting about lack of funds for this task. But think about the monetary loss in terms of lost production and avoidable litigation not counting the man hours wasted. An effective computerized MIS (management information system), integrated with GIS, will go a long way to introduce a semblance of transparency in forest department. Selected items on MIS and GIS can be made available on the websites of state forest department so that general public is also in touch with the activities. A feed back feature can be integrated with websites for two-way communication and proper dialogue. Field level officers should be equipped with digital video cameras. Instead of inspection notes, video films of field activities will better illustrate the state of affairs. The videos can be transferred on CD and circulated at appropriate levels. Remote sensing is another frontier technology with vast potential that has been overlooked by state forest departments. It is a process of gathering information about an area, an object or a process without being in contact with it. Most often, remote sensing conjures up an image of satellites orbiting in space. However, remote sensing can be practiced at many scales. Resolution, the smallest unit that can be detected, is related to the altitude of the sensor. For instance, Thematic Mapper data from the Landsat satellite is sensed from an altitude of 705 km with a resolution of 30m, while aerial photography from an altitude of 460 m will yield digital imagery with a working resolution of 0.50 m. For even better resolution, a stable platform operating at a lower altitude is necessary. A balloon (blimp) fitted with camera can yield remarkable results. In an actual case study, a blimp 5.5 m X 2 m with a capacity of 9 m^3 of helium was used. It produces a net lift of 4.7 kg and maximum altitude attainable is 152 m. A custom-made gondola holds two 35 mm cameras positioned to photograph the same scene. The gondola contains two radio-controlled servos, one attached to a dual shutter release so that both cameras fire simultaneously, and the second attached to a stepping motor that rotates the gondola to align shots. The blimp is controlled by a tether. Cameras are autofocus, autowind 35 mm single lens reflex type fitted with 28 mm lenses. Filters are attached to lenses to match film types. In this case study, IDRISI remote sensing/GIS software process images and data because this is raster based and can be used with PC. It is powerful yet easy to learn and relatively inexpensive. A Trimble Pathfinder Geo-Positioning (GPS) system was used to locate the trees and fence positions for referencing to Universal Transverse Mercator (UTM) coordinate grid.The positions were differentially corrected to be accurate within 2 m. All corrected points were brought into AutoCad as Drawing Interchange Format (DXF) files for final fitting of an ideal tree grid. The best-fit grid was used to supply tree coordinates for processing of aerial photos. The final image gave an estimate of 6.6% for tree canopy cover. The error was only 5% as a result of comparison with ground reality. Remotely-sensed images can be linked to relational database containing trees and site statistics.

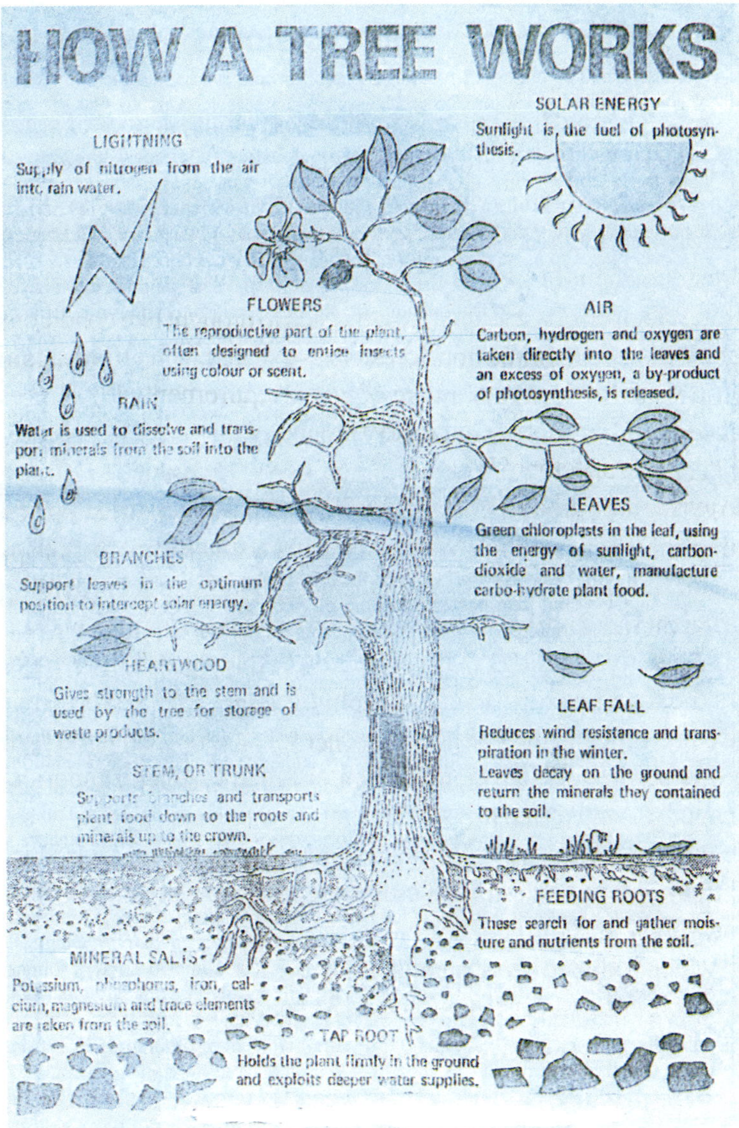
HOW A TREE WORKS
SOLAR ENERGY
Sunlight is the fuel of photosynthesis.
LIGHTNING
Supply of nitrogen from the air into rain water.
FLOWERS
The reproductive part of the plant, often designed to entice insects using colour or scent.
AIR
Carbon, hydrogen and oxygen are taken directly into the leaves and an excess of oxygen, a by-product of photosynthesis, is released.
RAIN
Water is used to dissolve and transport minerals from the soil into the plant.
LEAVES
Green chloroplasts in the leaf, using the energy of sunlight, carbon-dioxide and water, manufacture carbo-hydrate plant food.
BRANCHES
Support leaves in the optimum position to intercept solar energy.
HEARTWOOD
Gives strength to the stem and is used by the tree for storage of waste products.
LEAF FALL
Reduces wind resistance and transpiration in the winter.
Leaves decay on the ground and return the minerals they contained to the soil.
STEM, OR TRUNK
Supports branches and transports plant food down to the roots and minerals up to the crown.
FEEDING ROOTS
These search for and gather moisture and nutrients from the soil.
MINERAL SALTS
Potassium, phosphorus, iron, calcium, magnesium and trace elements are taken from the soil.
TAP ROOT
Holds the plant firmly in the ground and exploits deeper water supplies.

2 Water For Life

All literate Indians must have read the famous lines of noted Hindi Poet Rahim –" रहिमन पानी राखिए बिन पानी सब सून " i.e. *water is the essence of all life forms.* Nature's boon in the form of water is not free of cost and unlimited. Modern day foresters of India pay scant regard to importance of water in afforestation efforts. Failures are attributed to drought like conditions during monsoon. Some non-foresters associated with afforestation effort were instinctively aware of water requirement of saplings planted in difficult tracts during dry months of the year. For instance, Fishcer was collector of Etawah during 1890s. He wanted to afforest the ravine area around Etawah city. First step was construction of a dam in one of the gullies. Water stored in the catchment on the up-stream side percolated slowly in the neighbouring tract providing sufficient moisture to the Sisso (Dalbergia Sisso) saplings planted subsequently resulting in lush plantations. Alas, the dam is no more and the tract has reverted to xerophytic stage. The Sissoo trees have been cut and the legendary Fischer forest is now covered with thorny scrub vegetation. Mr. Allen of Allen Cooper Tanneries, Kanpur was interested in a sylvan retreat overlooking a water body. He purchased a tract of barren ravenous land on the outskirts of Kanpur and put a dam in one of the gullies. On the edge of the water body thus formed, he constructed a lodge. Percolation from the water body provided enough moisture for the plantation in the neighbouring tract which now houses Kanpur Zoo and the residential colony of forest officials.

The message is – try to create water bodies wherever feasible before undertaking plantation. Survival will be higher during critical summer months and growth will be throughout the year instead of restricted only to monsoon season and during winter rains. Higher survival rate and shorter rotation for same volume growth will more than justify the additional expenditure. With the advent of canal system in west Punjab (now Pakistan) the forest department came up with irrigated mixed plantations of Sissoo and Mulberry in Changa Manga. These proved to be phenomenally successful and contributed significantly to the sports goods industry of Sialkot. After independence, Haryana forest department replicated the concept at Saraswati – though the irrigation was through tube wells. Tube-well irrigation has drawbacks like lowering of water table and transformation of fertile areas to saline alkaline ones. In fact, flood irrigation is not required in any form. Vegetation is a self-regenerating system which creates own reservoir of required nutrients. A prime example can be observed in Panki Thermal Power Plant at Kanpur. In an experiment for stabilization of fly ash, leguminous tree species were planted with no other inputs except irrigation by waste water from the thermal power plant, of course, after cooling. Flyash is a totally inert matter devoid of all organic ingredients capable of sustaining life. Flyash dump provided only the anchor for saplings which obtained necessary moisture from irrigation provided. Sunlight, carbondioxide and moisture produced the necessary food. Nitrogenous nutrients were provided by nodules in the root system. Seeds of grasses and herbs were brought in by wind, insects and birds and germinated in the now fertile soil. Other micronutrients were provided by dead insects and droppings of birds. Within a short span of few years, the flyash dump turned into a lush green, self supporting eco system harbouring birds, rodents and reptiles as well as umpteen types of insects. Now it has been proved that apart from leguminous plants, other species also are associated with mycorhiza which can fix Potassium, Phosphorus etc. To legislate is one thing and to regulate another. Flyash notification of Union Ministry of Environment and Forests (MEF) states that by September 2002, old power plants should be able to show 20% flyash utilization. Further, manufacture of clay and tiles is banned within

50 kilometre radius of thermal power plants without at least 25% mixture of flyash. Four years down the line, targets are illusive. Currently 90 million tonnes of flyash is being generated annually in India. By 2012 the figure is projected to rise to 200 million. Already 32,000 hectares of land is used as ash ponds. At present only 13% of flyash is being used for manufacture of tiles, bricks and cement.

Vegetation requires moisture in miniscule quantities which can be made available by any of the following means.

1. Maintaining enough humus in the soil by not removing the dead vegetational residue from the system and allowing it to decay in situ. Humus has a property to conserve soil moisture and provide essential nutrients. Alternatively, if the soil is degraded and devoid of vegetation, humus can be added with inputs of compost, vermicompost or organic manure which are now available commercially.

2. Mulching is an excellent method of conserving soil moisture and is beneficial to plants because –

    - Smaller area is exposed to wind and sun
    - Rainfall is concentrated onto smaller area so that it percolates deeper into soil and is less exposed to the forces causing evaporation losses
    - Water absorption is accelerated by preventing raindrops from compacting a large part of soil surface
    - Soil is conserved
    - There is a moderating effect on soil temperature
    - Reduction in soil salinity
    - Weeds are kept in control
    - Soil structure is improved
    - Nutrient status of soil is improved

Simplest method of mulching is by hoeing. In fact, several weedings and hoeings are prescribed in many plantation techniques. As indicated in an earlier chapter, forest officials avoid carrying out some of the prescribed operations due to avarice resulting in retarded growth and mortality. Mulching with dry vegetation is not advised in tropical countries due to possibility of termite infestation. Mulching with pebbles is excellent – though it is not popular in India. Mulching film made of polythene is a very efficient method of soil moisture conservation as well as weed control. These are available in several versions like super thin, dual colour etc. for specific uses. In addition, in sloping tracts, it can be useful for soil conservation. The only drawbacks are possibility of damage and theft by antisocial elements. In areas under JFM, it should not be too difficult to obviate such damages by putting the onus of safeguarding losses on local inhabitants. Major features of mulch film are –

- That it insulates the plant from moist soil preventing contact reducing rot.
- In drier micro climates, it reduces plant sensitivity to disease.
- It also reduces the incidence of insects and insect-vectored viral diseases.
- It prevents water loss due to evaporation and maintains humidity in the root zone.
- It prevents weed germination and growth.
- It adjusts light and heat conditions and provides the optimal root-zone temperature.

3. Another cheap method can be implanting an earthen pitcher by the side of the sapling. A tiny hole is drilled at the bottom of the pitcher. A rag spun in the shape of a string is inserted in the hole. If the pitcher is periodically filled with water, enough

moisture is maintained in the root zone. In JFM areas the beneficiary should do the job of filling the pitcher once in ten days during dry spells. In other areas, the paid watchman can undertake this task on a rotational basis. This model can be observed in areas around Chandi Devi temple near Hardwar.

4. If small water bodies are formed scattered all over the plantation area, these can be useful for sprinkler, drip or mist irrigation. The water is to be stored through rainwater harvesting. Losses due to evaporation can be minimized by addition of chemicals like hexadecanol and other long-chain fatty alcohols which form a thin monomolecular film on water surface. Another method is to float a special type of cloth or hollow concrete blocks or Styrofoam blocks on the water surface. With JFM dispensation, it should not be too difficult to prevent thefts of these materials. Losses due to percolation can be prevented by painting the bottom of reservoir with polyurathene paint. The reservoir can be natural depressions or dug-up shallow ditches. Such reservoirs can be useful in forest fire prevention by misting and for wildlife conservation during dry spell. The techniques have been described at appropriate places elsewhere in the book. Rain guns can be very efficient means of providing moisture to vegetation from water bodies. One such product of Italian origin is illustrated in Fig. 1 below. The unique advantage of rain gun is that it can discharge high pressure water rotating at a uniform speed over various diameters. This way the water is distributed all over the irrigated area. International Development Enterprises (IDE) is headquartered at USA and has its India office in New Delhi. It is a non-profit development organisation with a mission to develop and promote affordable technologies to increase the income of small and marginal farmers. IDE has successfully developed various affordable micro-irrigation

Ultra Sonic Mist Generator

Rain gun

Rain gun in action

technologies. Micro irrigation increases yield to the tune of 20% and saves water from 30% to 80%. Crops mature earlier and fetch better prices. Costs on inter-culture, labour etc are reduced. Undulating terrain, saline, water-logged, sandy and hilly lands can be brought under efficient cultivation. Water-soluble liquid fertilizer can be applied along with irrigation. At the other end of the spectrum there are sophisticated computerized irrigation and fertigation systems of Israeli origin. Water filtration devices are in built to prevent clogging. Such systems supply the exact amount of NPK+ micronutrients on a continuous balanced feeding throughout the cycle resulting in optimum plant growth. Such system controls the Electrical Conductivity and pH value of water as well apart from humidity and temperature control. Labour and water saving is up to 90%. Such systems are for multi-tier plantations being advocated in subsequent chapters. **Tractor mounted sprayers** are very versatile, economic and effective. Both ground spray boom and overhead spray boom attachments are available. A 400 litre capacity fibre glass tank is attached with the sprayer. Thus, with less manpower, a large area is covered. **Drippers, mini sprinklers** and **foggers** along with compression fittings and sand filters are also in the market. All purpose turbulent flow clog resistant openable drippers are very user friendly. Another innovative contraption is **subsoil injector** which helps to water the roots up to 9 inches below ground. It results in huge savings on water, fertilizers, pesticides and boosting growth and yield of vegetation. The product is an enlarged version of a medical syringe, operated by foot. It is fed by gravity from a 20 litre can carried either on the back of operator or mounted on a vehicle. It is easy to use. The needle is placed near the sapling at an angle of about 45° and forced into the ground in one push. The pedal is then pressed with the foot. This operates a piston that forces out up to one litre of water at nine inches which is the ideal depth of optimum utilization. The Rubber Board (Kottayam), ICAR (New Delhi) and UPASI (Coonoor) are recommending this innovative solution. The advantages of this technique are many –

- There is saving of water
- No evaporation losses
- Moisture is retained for a longer time
- No weed growth as the surface is not wetted

5. Chemical water surfactants are now available. If mixed with water and sprayed on soil, water molecules develop adhesive qualities and stick to soil particles for weeks together maintaining soil moisture. Good quality portable sprayers and mist generators are now available. Small quantity of water can be taken from ponds described in para 4 above. Alternatively, misting can be done on leaves. Water molecules will stick to surface of leaves and will be slowly absorbed to offset water stress of saplings. Neem (*Azadirachta indica*)-based wetting agents are also being marketed. One such product is Agrowet.

6. Ultrasonic fog generators (UFG) have ushered in agriculture of tomorrow. The UFG fog behaves as a fluid in gaseous phase which is capable of deep penetration of minute spaces. This eliminates the free water which contributes to contamination, disease and pollution. In addition it replaces the use of chemicals in effective dispersal of a broad range of treatment solutions. UFG is a low maintenance combination of solid state electronics and modular construction of few moving parts with no small aperture or pressure requirements. One of the characteristics of the ingenious aeroponic system is that it provides plant growth and root support with or

without a soil bed, the plant and root system being fully engulfed by an ultra fine (5 micron) nutrient vapour.

7. A rainfall simulation system to generate artificial rainfall has been developed in the Department of Soil and Water Conservation Engineering, G. B. Pant University of Agriculture and Technology, Pantnagar. It consists of a slope adjusting hydraulic flume of 10m X 1.2m X 0.5m size and a water circulation and regulation system. Hypodermic needles are used and it covers an area of 10m X 1.4m. It produces rainfall in varying intensity which is very similar to natural rainfall. The slope of hydraulic flume can be adjusted manually and through a motorized mechanism between 0% to 6%. The water circulation system is capable of regulating flow up to 100 litres per second.

8. During a period of six months from October to March, dew accumulation in the belt stretching from Punjab to Assam is 15 to 30mm. About 25% to 50% water deficiency in this belt can be met by harvesting dew. The technique is to cultivate the field in the evening and then planking in morning hours. Mulching with pebbles also has been used for dew harvesting.

9. In sandy areas, horizontal asphalt barrier 55 to 60cm below soil surface may appreciably increase water available to plants.

3 Let Us Turn To That Lowly Worm

People in the age group of above fifty must be nostalgically remembering the earthworms wriggling in thousands during monsoon season. The sight became rarer with passage of time. The green revolution with attendant chemical fertilizers, insecticides, pesticides, fungicides and weedicides must have taken their toll. There are more than 500 species of earthworms in India. Earthworms can consume practically all kinds of organic residues. They consume organic residue which can be as much as 2 to 5 times their body weight. Only 5% to 10% of this ingested matter is used as food stock for body growth and the undigested matter is excreted as wormcasts which is mucus coated. Wormcast consists of organic matter that has undergone physical and chemical breakdown through the activity of muscular gizzard which grinds the material to a particle size of 1 to 2 micron. Wormcast is a rich source of macro and micro nutrients, vitamins, enzymes, antibiotics, growth hormones and immobilized microflora. The nutrients present in wormcasts are mostly in available form. Broadly, earthworms are classified into three categories.

- Epigiecs i.e. surface dwellers e.g. Eudrilus eugeniae, Eisenia foetida and Perionyx excavatus which are exotics. Indigenous species are Lampito mauritii, Octochaetona serrata and Perionyx excavatus.
- Subsurface dwellers e.g. Lampito mauritii
- Endogeics e.g. Octochaetona thurstonij

Some work has been done on Epigiec worms in production vermicompost. This practice has to be made widespread. It will have a threefold impact. Firstly, afforestation areas and agricultural areas will get excellent organic humus with lots of micronutrients which will increase productivity. Secondly, dependence on chemical fertilizers will reduce offsetting pollution. Thirdly, it will get rid of mountains of domestic refuse which are creating disposal problems apart from being health hazards. The method is very simple. In households and eating establishments, two separate receptacles can be kept – one for biodegradable waste and the other for non-biodegradable waste. It should not be too difficult to devise mechanical contraptions to handle the receptacles in order to maintain hygienic conditions. The non-biodegradable portion can be recycled. The biodegradable portion can be put in pits or troughs, alternate layers being interspersed with Epigiec worms. Within a short time, the refuse will turn into vermicompost which can be commercially marketed or utilized by forest departments in afforestation areas. Vermicompost influence the physical, chemical and biological properties of soil.

- It improves the water holding capacity of soil. The mucus associated with wormcast is hygroscopic and prevents water logging.
- It acts as a buffer in maintaining soil pH.
- It influences the availability of micronutrients by chelating them as complexes
- It possesses high CEC and phosphate solubilising actinomycetes which in turn influence the soil fertility.
- It possesses vitamins and growth hormones which have a direct role on plant growth.
- It acts as a mine for various plant essential nutrients such as N,P,K,S and trace elements.
- It increases porosity of soil by mixing top soil and sub soil.
- Due to multiplication of earthworms, incidence of nematodes is reduced.

- By increasing population of micro and macro organisms organic residue decomposition is accelerated.
- Helps in controlling soil erosion and soil degradation.
- Reduces nutrient losses due to leaching.

It is a pity that forest departments have not taken a lead in the matter. At least in the residential colonies housing forest personnel, the practice of vermicomposting can be enforced. There is a lesson for municipal corporations like Muradabad and Kanpur as well. All types of refuse are dumped on the outskirts of the cities at designated dumping grounds. The visitor to the city is welcomed with an offensive smell and inhabitants are subjected to health hazards. The counter argument will be non availability of funds. Remember the plague scare from Surat a few years back and consequent loss to the national exchequer not counting the human lives lost. After all, there is little value attached to human lives in the present state of affairs. Critics of vermicomposting cite some disadvantages like accumulation of heavy metals like cadmium, nickel, chromium, zinc, copper etc. which can adversely affect the food chain. But advantages far outweigh disadvantages. The recycling of organic waste can be done by

- Bed method:- It is a simple process and can be easily practiced by any farmer. A tank of 2mX1mX1m needs to be constructed in a shady location. At the bottom of the tank, a layer of stones and sand particles have to be put which would facilitate the drainage of water. Above this a 15cm layer of loamy soil has to be put and sprayed with 4-5kg of diluted cowdung. This layer acts as a bed for earthworms. In this desired species of earthworm (about 100 in number) like Eudrilus eugeniae, Eisenia foetida, Perinyx excavatus may be placed. Above this a 5cm layer of organic waste has to be added. This should be kept moist by spraying some water on it. Every fourth day organic waste have to be added till the pit is full. At that stage cowdung slurry has to be sprayed and the pit covered with gunny bag cloth. The bed has to be kept moist by off and on spraying. After 25 days the entire mass has to be turned upside down to induce movement of worms. After 2-3 months, the material will be ready as compost. At this juncture, moistening should be stopped to ensure drying of compost. The compost formed is of grey to brown coloured loose mass. The nature of vermicompost is influenced by the type of organic waste and worm used.

- Commercial method :- A commercial vermicomposting production plant requires special technical and management skills. It requires large quantities of organic waste residues at assured low price for the project period (5 to 10 years). For a commercially viable unit, the daily production capacity has to be 5 to 25 tonnes per day depending on the availability and cost of raw material delivered at the site. The commercial plants will be able to market the vermicastings to the farmers, nurseries and households. Depending on the cost of raw material, the project cost is recovered within 1 to 2 years. Press mud, waste products of fruit and vegetable processing units and dairy or poultry wastes as well as crop residue can be used. Waste is heaped in rows and the waste is treated with microbial culture for defouling and accelerated bio-conversion. The required moisture is maintained by spraying with water. The waste is left to decompose. During this period, the decomposing organisms multiply and in the process consume more oxygen to break down the organic compound and release heat energy through respiration process which causes the temperature to rise in the decomposing matter. Temperature of the heap can reach 75° C killing pathogens and ensuring proper composting. The heap is churned frequently to provide sufficient oxygen for aerobic composting. After two weeks, the decomposing organic matter is ready for vermicomposting. Organic wastes are separated from non-biodegradable matter. Optimum conditions of temperature, moisture and pH are maintained for speedy bio-conversion in an ecofriendly manner. The vermicastings are then produced by specially selected breeds of earthworms which carry an aerobic bioconversion of diverse organic matter in an ecofriendly manner. Vermicastings are harvested periodically and sieved. The vermicastings are then blended with formulations, standardized and packed for sale. It can be blended with bio fertilizers containing nitrogen fixing and phosphate solubilising bacteria to ensure bio availability of nutrients.

Endogeic worms require more research regarding their breeding in laboratories. The bred worms can be released in afforestation sites. The tunnels will improve aeration and percolation. The castings of the worms and mucus of their bodies provide excellent nutrients for vegetation. The worms themselves attract predators which in turn improves the ecological status.

Nature uses a little but gives a lot

4 Resort To External Inputs

The emphasis in Indian agriculture (and forestry) has been more on exploitation of natural resources – land and water – but negligible on building up, restoring, reclaiming and enhancing their productivity and sustainability. It is essential to make efforts for the improvement of land, water, solar energy and biotic resources management practices. To make the system sustainable, it is imperative that economic development today should not be at the expense of prospects tomorrow. Modern agriculture lays too much emphasis on synthetic inputs for spectacular results. We should endevour to strike a balance between high productivity and sustainability. So called waste lands – or wasted land – offers us ground for trying out state-of-the-art techniques because there is nothing further to loose. The latest thinking is that land only offers an anchorage for roots of vegetation. All other items are available in the ecological system or can be supplied as additional external inputs. Organic farming is also gaining momentum and in monetary terms, can be viable proposition as affluent health-conscious people now are willing to pay higher price for organic food and it has good export potential. Organic food is more palatable as well. There are international organizations to certify organically grown items so that there is quality control. Development of biological cycles involving micro organisms, soil fauna, plants and animals is the basis of organic agriculture. Switzerland-based Research Institute of Organic Agriculture shows after a 21-year study that organic approach more than makes up the difference in crop yields through its ecological benefits. By statistically designed experiments it was proved that organic systems were able to produce more with less energy and fewer resources. Thirty four percent to fifty one percent less nitrogen, phosphorus and other nutrients were added to the soil than in conventional systems. In spite of this crop yield was just 20% lower which proves that organic systems use resources more efficiently. The organic soils were also found to be better in other key ways:

- A larger and more diverse community of organisms were hosted
- A larger and more diverse community of microbes was hosted
- Organic farmers help in increasing biodiversity

Many items are available commercially which can be beneficial for plants. The only problem is that many unscrupulous venders, hand in glove with officials, try to fob off spurious or substandard items. Naturally, the desired results are not obtained giving a bad name to innovative ideas. This brings us back to what has been narrated in preface and first chapter.

Plant growth promoters (PGP) are already being used in agriculture. If we are take up forestry and agro forestry as a serious commercial enterprise, then possibility of application of PGP has to be explored. Sometimes plants refuse to respond in spite of doing all the right things. It is because the plant does not release its hormones during the crucial vegetative/reproductive phase. PGP improves the physiological efficiency of plants, promotes growth and development, leading to higher yields and better success. PGP is a mixture of long chain high molecular weight fatty alcohol with carbon atoms ranging from 24 to 38. Besides there are sterols, some of which are identified as â-sitosterol, campa sterol and sigma sterol. PGP is neither a fertilizer, nor a fungicide, nor a pesticide nor a micronutrient mixture. PGP is taken up by plants through the cuticular layer and through epidermal and endo-dermal cells. It is also taken up by root cells when applied to soil. After PGP is absorbed in the plant, the active principles in it activate several enzymes or biological catalysts in plant tissues. PGP also encourages the roots of the plant to penetrate deeper into the soil and absorb more nutrients. Due to

PGP spray, the cell wall becomes more permeable to absorb and translocate more nutrients and solutes and ensures effective utilization of nutrients absorbed by the plants. All these result in increased plant size, better health and more production. The right proportion for most commercially available items is 100ml PGP and 200ml of water for foliar spray. The quantity can be suitably diluted for spraying on soil. If water surfactant is also added then amount is further reduced. Most commercially available PGPs are safe for humans and animals. Granules are also available for soil application in case spraying is not feasible. In seed production areas, PGP can be sprayed 10 to 15 days before flowering followed by applications after 30 days and 60 days. Application is useful in nurseries as well. Water soluble fertilizers being marketed by a firm in Maharashtra is noteworthy because it does not increase salinity of water and soil. These are acidic in nature. The acidic solution keeps dripper/ sprinkler lines and nozzles clean saving cost of de-clogging. There is quick absorption by roots and there is negligible leaching, soil fixation and evaporation losses of nutrients. Very low concentration solutions (0.1 to 1%) are used which ensure use of exact quantity of nutrients as per requirements of the plants, thus avoiding wastage.

Knitted shade cloth is a very useful and versatile item in the market. The advantages are as under:

- For shade control three shade percentages are available- 25%, 50% and 75%. This enables the user to control the intensity of sunlight with even dispersion below the shade.
- It is light weight making installation and removal easy.
- It is durable being UV stabilized for long life.
- It can be removed, washed and reinstalled.
- Being antifungal, does not require germicidal treatment. That way it is superior to coir matting requiring antifungal spraying.

- Being non hygroscopic, it does not absorb water. Hence it remains easy to handle during monsoons. On the other hand, coir or jute sags under its own weight during monsoon.
- Being air permeable, permits even circulation of air in the covered area with no need for extra air vents (as is required with fixed conventional shades).
- It can be used for protection of nurseries and high value crops from excessive sunlight, wind and hail.
- It can be used for collection of seed and fruit.
- It can be used as anti-evaporation net for irrigation reservoirs and open water tanks as the fabric floats on water.
- It can be used for drying seed and fruit.
- It can be used for shading and as windbreaker.
- Since we are talking in terms of integrated land use pattern, it can be used for shading of livestock and as animal sheds.
- It offers quick relief in various situations at an economical cost.
- Flexibility is offered with various constructions to suit individual applications. Instantaneous clips are available for ease of installation of the fabric. The clips can be put anywhere, are strong, effective as they connect a large surface area, are reusable, UV protected for longer life. Number of clips can used as per the intensity of wind expected.
- The net is available in standard size of 3mX50m in black and green colour.

One Israeli firm has introduced "Sun Selector Green House Covers" to help plants grow better. These are highly advanced, long lasting, multilayer photo selective films used in greenhouse and tunnel covering applications. These films combine high-level agronomic research with quarter of a century of professional plastic know-how to provide growers with films that are technically unique and economically viable. The unique mechanical, optical and thermal characteristics of this film ensure greater durability, excellent resistance to weathering and reduced pest activity.

For meeting bourgeoning demands, use of pesticides has registered an upward trend. Since foresters are per-se environmentalists as well, they should resort to eco friendly technologies for pest management instead of chemical pesticides because of their easy availability and immediate spectacular effect. Pesticides which have been banned are listed in Table I.

Table I

1.	Aldrin	11.	Copper acetoarsenite
2.	Calcium cyanide	12.	Dibromochloropropanes
3.	Chlordane	13.	Ethyl Mercury Chloride
4.	Endrin	14.	Ethyl parathion
5.	Menaron	15.	Nicotine sulphate
6.	Nitrofen	16.	Paraquare dimethylsulphate
7.	Heptachlor	17.	Pentachloro nitrobenzene
8.	Tetradifon	18.	Pentachlorophenol
9.	Toxaphene	19.	Phenyl Mercury Acetate
10.	BHC	20.	Sodium methane arsenate

Following pesticides are for restricted use only.

- Aluminum phosphide can be used in rat burrows if packed in cage as tablet or pouch-in-pouch packing as powder weighing 3 gms each.

- BHC is banned for use in edible items.
- Captafol can be used only as a seed dresser.
- Carbaryl should not be sprayed on flowering stage.
- Chlorobenzilate can be used in control of mites of honey bees.
- DDT can be used only under supervision of government agency.
- Dieldrin can be used for locust control.
- Ethylene dibromide can be used as fumigant for foodgrain.

Pesticide regulations in India are lax. The industry has exploited the loopholes to corrupt the system and the government has turned a blind eye to the problem. Government must put people's health above everything else. If we are to shun chemical inputs, then what are the alternatives available? Organic manure prepared from bone meal is one such item. It sustains the fertility of soil and improves the moisture retaining capacity of even sandy soils. On the other hand, it helps in reducing moisture content of black soils. It improves the aeration of soil and maintains a congenial environment for useful micro flora and micro fauna. If used in conjunction with fertilizers, the efficacy is enhanced, warding off the ill effects. It helps the plants to absorb nitrogen and other nutrients. It helps in neutralizing alkalinity of mildly *usar* areas. Neem-based organic manures work as pesticides as well. Neem-based urea coating agents help in reducing nitrogen losses thus saving 50% cost of urea improving productivity due to prolonged and continuous availability of nitrogen to plants. Neem-based pesticides repel insects due to the bitterness imparted to plants which render the edible parts unpalatable for insects. Bio fertilizer liquid and granules help the plant in critical stress period. Healthier plants show more resistance to drought, frost and disease. Crop quality and yield improve which can be stored for longer periods. Some firms are marketing plant food liquid which are systemic type of essential nutrient that can be absorbed by the plant. Non-toxic, biodegradable and eco friendly seed coating powder protects seed from virus, fungus and bacteria. Then there are zeolite containing soil conditioners which work within hours with lesser quantity compared to gypsum and bring pH of soil to normal. These prevent hardening of soil due to injudicious use of chemical fertilizers. Also available are di-carbooxylic group-based translocaters. The molecules of this chemical carry the food prepared in the plant to the fastest growing tips in right proportion.

Weeds have always been a matter of great concern. They compete with the preferred species for light, nutrients, moisture and space. Though herbicides may have an adverse effect on environment, sometimes there may be compelling reasons for their use. Some commonly used herbicides are

- Atrazin: It is a selective herbicide used as preemergence control of annual grasses and broadleaved weeds.
- Alachlor: Used for preemergence control of annual grasses, broad leave weeds and sedges.
- Diuron: It is a selective herbicide used as preemergence control of annual grasses and broadleaved weeds.
- Metribuzin: Used for post emergence control of grasses.
- Glyphosate: It is nonselective.
- Paraquat: It is also nonselective.
- 2,4-D: Used for post emergence control of broad leaved weeds.

Pheromone traps are very effective in controlling insect pests. These are used for luring male moths. These can also be used as a confusing strategy or mating disruption technique. These traps are highly species specific. In simple terms, pheromones are hormones secreted by females to invite males for mating. Scientists have identified the chemical signals which attract male insects, and have succeeded in using this principle to attract male insect in good numbers for effectively managing their population. After deciphering the chemical structure of these pheromones, synthetic substitutes have been produced. Simple light traps are also being used extensively as part of the integrated pest management strategies for various crops. In rice cum fish culture, the light traps are used for attracting the insects which fall in water below and become feed for fish.

5 Poor Man's Timber Can Alleviate Poverty

Some members of *Gramineae* family play a very crucial role in the economic life of the nation. Bamboo is one important member of this family. Bamboos are one of the most important renewable forest produce which has a capability to produce maximum biomass per unit area per unit time as compared to other forest species. One hundred and thirty eight species of bamboo are found in India. There are about 1500 documented traditional uses of bamboo from cradle to coffin. By virtue of its availability and versatility, it is associated with the traditional rural lifestyle. Bamboo has been rightly given the sobriquet of poor man's timber. It is extensively used for construction, ladders, small bridges, culverts, fences, furniture, agricultural implements, cots, walking and other types of sticks, tent poles, kites, brush, musical instruments, fans, bows, arrows, lances, baskets, boxes, mats, curtains, blinds, umbrellas, fishing rods, cradles, candy, chutney, curry, vinegar or lime pickles, curries, *tabashir, banslochan, fodder etc.* – the list is ad infinitum. Young shoots of both running and clump forming bamboos are used for edible purpose. Traditionally, edible bamboos have formed a part of the human diet particularly by tribals of north-eastern part of India. Tender shoots of *Dendrocalamus giganteus, D. hamiltonii, D. membranaceus, D. strictus etc.* are consumed as vegetable or pickles. In Manipur, *Bamboosa bamboos and D. giganteus* are the two main species whose young shoots are used as vegetables. Tender shoots of *B. bamboos* are pickled and steeped in oil for direct consumption or made into curries. *B. polymorpha* shoots have a distinct sweet taste and is considered to be the best the world over for producing quality edible shoots. The crooked tender shoots of *b. balcoa* are eaten and preserved also after fermenting and drying. Bamboo seeds are pickled and used for making beer. The leaves of a number of bamboo species are much valued as fodder during scarcity. Young bamboo leaves are favourite fodder of elephants. Fruits of *Melocanna baccifera* are devoured by cattle, elephants, pigs etc. In India, seeds of *B. arundinacea, D. strictus and M. baccifera* are extensively eaten by tribals and poor during famine. Most of the species growing in Thailand produce edible shoots, the best ones being *D. asper, D. giganteus, D. merrillanus, B. tulda, Pseudoxytenathera albociliata and Thyrostachys siamensis.* Shoots of bamboo species are consumed in fresh, pickled and dried forms. In Japan more than 8000 tonnes of bamboo shoots are consumed every year, the main species being *Phyllostachys pubescens.* About 400 tonnes of young shoots of *P. pubescens* are canned and exported to Japan from Thailand. In Taiwan the main edible species are *P. edulis* and *D. latiflorus.* Management of these species have reached a high degree of specialization there. About 10,000 kg of bamboo shoots are obtained from one hectare of intensively managed area in Korea. Healthy shoots of bamboo, stunted or even crooked are used for preparation of edible products. The outer leaves or sheaths are removed and the tender portions are cut into rings or pieces of suitable sizes. Bamboo shoots contain cyanogenic glucocide which on endogenic hydrolysis yield hydrocyanic acid. Hydrocyanic acid can be removed by steaming the pieces of bamboo shoots. Another toxic constituent, taxiphylin is found in the shoots of *B. guadua, B. vulgaris, B. arundinacea, D. giganteus, and D. hamiltonii.* This toxin can be removed during cooking the shoots by changing the water several times. Soaking thin pieces of shoots in different changes of 2% salt water for several hours also removes the toxicity. In many south eastern countries, the powder of tender unopened leaves of bamboo is used for brewing tea.The nutritive value of bamboo shoots is high but differs from species to species and region to region. In one test carbohydrate content was 4.5%, protein 2.6%, fat 0.3% and ash 0.9%. Thiamine, niacin, vitamin C, sodium, potassium,

Electric Trimmer Can Be Useful In Cutting Crooked Culms From Congested Bamboo Chumps.

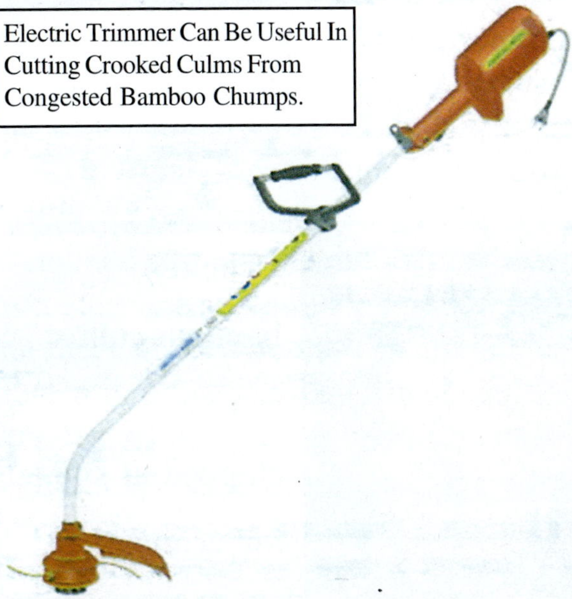

Crooked bamboo culms can be chipped at site for economical transportation.

calcium, magnesium and phosphorus were also detected. In Orrisa *B. bamboos seed* was found to be nutritionally equal to wheat but taste similar to rice.Bamboo is considered to be one of the oldest building materials used in rural areas. Bamboo scaffolding is also used in building construction and is extremely resilient and long lasting. Bamboo is used in the form of parquet, plybamboo and laminated bamboo. A number of bamboo species have tensile strength equivalent to that of iron and steel. Rice husk-bamboo mat board for concrete shuttering work is in vogue. Here the faces of the rice husk board is bonded with bamboo mats using phenolic resin. Bamboo fibre reinforced plastic (BFRP) holds great potential as the tensile strength of some BFRP composites has been found to be more than that of mild steel, while the density is only one eighth. Mechanical properties are similar to other commonly used composites such as glass fibre reinforced plastics. Grain silos have been manufactured from bamboo concrete. Durable bamboo laquerware is being manufactured in south east Asia. Main industrial use of bamboo is for pulp. These days, particle boards of bamboo are also coming into vogue. Apple packing cases have been designed and fabricated using two ply bamboo mat boards. Wirebound bamboo crates have been found to be better than poplar wirebound veneer boxes and are reusable. Culturally too, bamboos are close to the tribal people. Up to now, the main problem with bamboo plantations of most species on a large scale had been non availability of seed of desired species. Farmers propagate bamboo on a small scale through rhizomes. This was not feasible for large scale plantations by forest department. Bamboo flowers gregariously at long intervals, say forty years. So seeds have to be procured from far off places and even then supply is uncertain. It may not be phenotypically suitable for plantation site. Protection from animal – wild and domestic – is problematic. Result is congested clumps which are difficult to work by traditional methods of extraction. Output has no economic utility in the traditional commercial markets.

Yet, some lateral thinking for innovative ideas can turn the table. For instance, there are four miniature Taiwanese machines available which can be mounted on trucks and operated in situ, thereby saving the cost on transportation. These are for manufacture of bamboo tooth picks, ice cream spoons, Chinese chopsticks, incense sticks, blinds, chair, floor board and strips for laminated mats. Partially decayed bamboo and bamboo subjected to a period of weathering have been found useful for producing pencils. The congested and crocked culms can be clean cut with circular mechanized saws illustrated in Fig 5, leaving one or two culms per clump to supply food to the rhizome. The economic returns will make it worthwhile for forest corporations to put in an earthen mound round the clump for healthy future growth. Alternatively, clumps can be allocated to individual families for tending under JFM dispensation.

With advances in science, it is now possible to grow plants of many species of bamboo through tissue culture. The concept of tissue culture is born out of the fact that every cell in a plant is capable of regenerating into an entire plant, thus making it possible to regenerate innumerable number of plants from a single bud. The single nodal buds from identified phenotypes or auxiliary buds from seedlings/ young plants are taken as the basic material for culture. The time factor is critical which is when the bud just starts swelling. One month old seedlings can also provide explants from nodal portion. The defined protocol can be followed for generating propagules. This is a very specialized job requiring sophisticated infrastructure and is best got done through experts. The precaution to take is hardening of the tender plants before planting out in the field. Lack of this precaution have given poor results in the past giving a bad name to the technique. Another technique is **macroproliferation.** Four to five month old plants in the nursery developed through rhizomes develop good rhizomes of their own and have five to six tillers at this age. These tillers are separated along with a small piece of rhizome and roots. These plantlets attain the size of field plantable saplings within four months or these can be further multiplied through macroproliferation. By this technology, a large number of identified planting stock can be made available. Essentially the approach is same for all the species of bamboo viz. Dendrocalamus strictus, Bambusa arundinacea, Dendrocalamus asper (edible bamboo), Phyllostachys edulis etc. If one does not want to take the high-tech route and seeds are not available due to gregarious flowering at long intervals, vegetative propagation can be resorted to. Two to three year old culms are cut in such a way that there is a node in the middle. Cut portion is buried underground at a depth of 5cm in an horizontal position during February at interval of 30cm. Beds are prepared with mixture of sand, cowdung manure and farm yard manure. Watering is done at regular intervals. Within a month, shoots start sprouting and are given antifungal treatment. During June-July, the pieces are dug up with roots and planted out in the field. Rhizome planting is preferred by farmers who require only a few plants. In this technique, 30cm part of rhizome is dug up ensuring that one culm is attached to the rhizome before onset of monsoon and planted deep at site. If necessary, irrigation is done. Recent research has indicated that more than one culm may sprout from juvenile seedlings grown in amended soil. These culms are separated along with rhizome and roots and transplanted in polythene bags. Seed sources have also been developed from out of phase and early flowering varieties and induced flowering.

For block planting, spacing is 6m X 6m and fr planting on bunds the distance should be 6m. After 6 to 7 years the plant takes the shape of clump and three year or older culms can be extracted. About 6 culms should be left to sustain the clump on the outer periphery. Immature culms of less than one year in age should not be extracted. The stump should not be more than 30cm in height. In congested clumps, the culms causing congestion should be removed. In immature culms, the twisted portion should be removed. No extraction should be done between July and October. Earth should be heaped on clump so that new culms are not affected by sunlight. Trees causing light shade should be retained as bamboo is a shade bearing species. Cutting cycle is three to four years. Bamboo should be protected against browsing. Normally, annual yield is ten culms per clump per year. At 6m X 6m spacing, there are 270 clumps per hectare. Even if 200 clumps are productive, 2000 sticks of bamboo are available per annum per hectare which comes to about 3 tons by weight. At the rate of Rs 8/- per stick, the financial yield is to the tune of Rs 16,000/-. The cost of establishment is only Rs 8,000/- per hectare and annual maintenance is Rs 500/-. The promise of Vesicular-Arbuscular Mycorrhiza (VAM) has attracted interest on their role in agrobiology. Utility of VAM in bamboo plantations holds great promise and goal-oriented research is warranted.

6 Natives V/s Exotics

Pendulum of public opinion has been swinging from one extreme to the other so far as choice of species for India is concerned. Some persons at decision making levels were so enamored of high performing exotics like Eucalyptus that ecological aspects were ignored. Persons at the other end of the spectrum reacted adversely to large scale Eucalyptus planting. Now, the debate is, more or less, academic because Eucalyptus has entrenched itself in India and the clock cannot be turned back. All the same, it is a pity that the natural miscellaneous forests in Udhamsinghnagar, Nainital and other places were clearfelled. These areas were subsequently planted up with Poplar and Eucalyptus by so-called cognoscenti who should have known better. At one time, tropical pines were hot favourites. The controversy about clearfelling of Shorea robusta forests of Bastar killed the idea. Ideas about Powlonia and edible bamboo are simmering in the backburner. On balance, the objective should be to favour natives in a country like India so well endowed with biodiversity. Here again, India is such a vast country with myriads of agroclimatic zones. The native of one nook can be the exotic of another corner. The distinction gets blurred in case of species that are entrenched in their new habitat for a century or more. Teak from Burma (now Myanmar) in Uttar Pradesh is one example. Prosopis juliflora was introduced in Sindh from Mexico. Now it is part and parcel of the landscape in states like Rajsthan, Uttar Pradesh, Madhya Pradesh, Delhi, Andhra Pradesh etc. Lantana and Parthenium are obnoxious weeds which defy all attempts of eradication. For Parthenium following measures can be adopted for eradication:

- In fallow lands manual removal can be tried.
- Chemicals like atrazine, 2,4-D sodium salt, bromacil sodium

chloride, paraquat and disodium methane arsenate can be used to kill strands of Parthenium.

- Pre-emergence application of simazine, diuron, metribuzin and terbutyn prevent weed growth upto 2-3 months.
- Spraying of 20% saline water and soap kills the weed effectively.
- The leguminous plant *Cassia uniflora* suppresses Parthenium due to allelopathic action. Other species like *Tephrosia purpuria, Amaranthus spinosa, Sida latifolia and Croton sparciflorus* smother Parthenium, thereby, diminishing its population.

All in all, the effort should be that indigenous species should get preference in afforestation and reforestation projects. In Uttaranchal, Rubus ellipticus grows in abundance. It is an evergreen shrub, with juicy fruits that contain a very high percentage of extractable juice and taste quite like raspberry. There is virtually no cultivation cost involved and is a good source of extra income for farmers. The fruits are rich in Vitamin C and Calcium. Wild apricot is a common fruit of the hills. This pulpy fruit is full of medicinal properties. The drawback is that it is not as sweet as the cultivated variety but good enough for chutney and jam. Wild apricots are full of protein. Figs are rich in Calcium, Iron and Vitamin C besides being delicious and fibrous. Punica granatum or wild pomegranate is economically important. Seeds are dried in sun to make *anardana* which imparts sour taste to chutneys and other preparations. This fruit is rich source of protein, Calcium, Phosphorus and Vitamin C. Wild fruit cultivation can be encouraged by planting in so-called wastelands or integrating these in agroforestry management systems. That way, overworked and undernourished women of hills can get some succour.

Amongst several deserts of the world, the scenario of Thar is very dismal. Population density is 61/km and livestock population is beyond carrying capacity. Inspite of the odds, people of this region have adapted themselves by identifying natural vegetation of the region which can sustain their lifestyle. These plants yield foodstuff which can be stored for long periods. Some such plants are being described as under.

1. **Khejri** or *Prosopis sineraria :* It is termed as *Kalptaru.* Pods are dried and subsequently used as vegetable. Fruit is called *Khokha* and eaten raw. Leaves are good fodder. Inflorescence is used as medicine.
2. **Ber** or *Zizyphus nummularia :* Fruits are eaten and leaves are good fodder.
3. **Ker** or *Capparis deciduas:* Half ripe fruit is used as vegetable and for making pickles.
4. **Kumat** or *Acacia Senegal :* Seeds are used as vegetable.
5. **Peelu** or *Salvadora oleoides* and *S. persica :* Fruits are eaten .
6. **Goonda** or *Cordia myxa:* Used as vegetable and for pickle.
7. **Goondi** or *C. garaf:* Used as vegetable and for pickle.
8. **Sahjan** or *Moringa oleifera:* Fruits are used as vegetable and leaves have medicinal value.
9. **Neem** or *Azadirachta indica:* Known for medicinal value the world over and fruits are eaten during scarcity.
10. **Jungal jalebi** or *Pithecellobium dulce:* Ripe fruits are eaten.
11. **Babool** or *Acacia nilotica:* Used as medicine and seeds are eaten as an alternative food.
12. **Phog** or *Caligonum polygonoides:* Flowers and buds are eaten with curd.

Wheat under poplar is a common practice in Uttranchal Tarai area

Laminated small timber can be strong as well as aesthetically pleasing in arches

13. **Imli** or *Tamarindus indica:* Fruit pulp is used as chutney. Leaves are added in vegetable preparations. Seeds are used as alternate food.

14. **Soma** or *Echnocloa colona:* Used as substitute of rice and eaten during fast as *sahgar.*

There is lot of hype about cultivation of medicinal plants in fields and wastelands. One view is that the medicinal value of a plant/ fauna can best be ensured if the species is collected from its natural habitat. This aspect needs research.

For making so-called wastelands or wasted lands productive, foresters will have to shed their mindset about planting only trees and ignoring shrubs, grasses and herbs. Suedia is a good species for reclaiming coastal and inland saline soils. *S. fruticosa*, *S. maritime* and *S. monoica* are the three species found in India. Parts of some species are used as human and veterinary medicine as well as fodder and vegetable. By repeatedly harvesting the crop, excess salt of the area is removed and the plant can tolerate pH value up to 8.5.

Subabool or *Leucaena leucocephala* attracted lot of attention during 1980s as a source of fodder. This tree was inducted from Phillipines and association of annual cereal and legume forage with Subabool can supply green fodder throughout the year. It has been reported that 5.05 t/ha per annum green forage production was obtained in one experiment when shorghum was intercropped with Subabool under rainfed condition in Jhansi. In other models Cowpea or Stylo were underplanted with Subabool.

Sooner than later the debate on exotics versus natives is going to be overtaken by that on genetically-engineered species. This will open a Pandora's Box for which foresters and environmentalists should be prepared well in advance. The best course will be not to shut the avenues of advance. At the same time genetically engineered plants should be restricted in total isolation so that gene pool is not polluted.

Fibre farms are an interesting option. Hybrid cottonwood can be grown on a rotation of six to eight years. In temperate regions *Robinia pseudacacia* (Black Locust) has done well. Organic matter and nitrogen (TKN) are the key nutrients of soil, which demonstrate fertility status and increases the availability of other nutrients for plant growth. In experiments, organic matter and TKN under Black Locust plantation system increased remarkably as compared to control.

Paulownia, a native of China has recently been introduced in India. It is a multipurpose tree valued for wood and non-wood forest produce. Based on the agroclimatic conditions in India, *Paulownia fortunei, P. tomentosa, P. fagesii, P. australis, and P.elogata* appear to have some potential in some regions of the country. The wood and bark have astringent properties. Seeds yield Abur oil used in preparation of paper. Leaves and flowers are rich in nitrogen and are good fodder. Charcoal manufactured from the wood is high class ingredient for gunpowder. Beautiful flowers make it suitable as an ornamental tree. Paulownia wood is good for furniture and carving. It is ideal for agroforestry. It can be propagated from seed or root cutting.ETPs may attain a height of 4-5 m in a year.

Many indigenous species have been planted extensively, but their economic potential has not been fully exploited. *Terminalia arjuna (Arjun)* is one such species which has been extensively planted not by design but by chance. In the hayday of social forestry, forest officials were desperately scouring for vacant land for utilization of overflowing funds and fulfilling annual targets. Most of the available land was subject to waterlogging. *Arjun* was one species suitable for waterlogged areas with stagnant water. Consequently, it became the preferred species for roadside, canalside and village common lands in eastern UP. The drainage pattern of this tract has undergone radical transformation due to construction of roads, canals and flood protection bunds. Rainwater, which used to flow to natural drainage channels, now is forced to stagnate for want of outlets. This has drastically altered the ecology of the tract inducting many diseases hitherto unknown – both for flora and fauna. Coming back to *Arjun,* hundreds of thousands of semi-mature trees remain unexploited. *Arjun* bark has following properties.

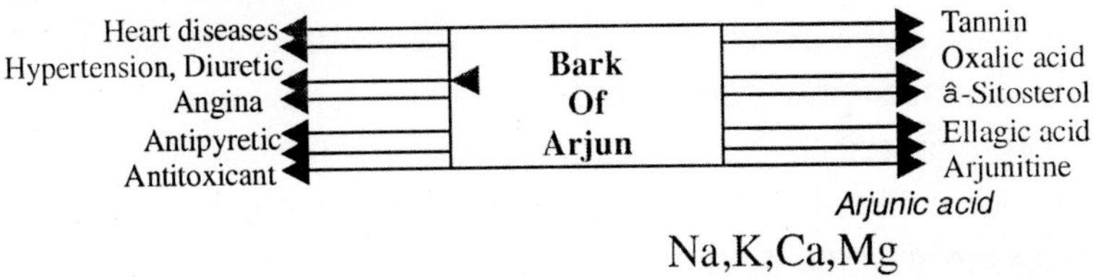

7 Breeding Tells

Forests are priceless reservoirs of plant germplasm and play a vital role in maintaining global environmental stability. The composition, structure and ecology of natural forests, with their large array of plant and animal species is very complex and still remains to be fully understood. Plants can be propagated through their two developmental life cycles – the sexual or the asexual. In the natural environment sexual and asexual reproduction has their appropriate selective advantages according to the stage of evolution of different kinds of plants. The genetic characteristics of planted trees are affected by choice of seed source and, in the long term, by breeding programme. The conservation of the diversity of forest genetic resources is a cornerstone for future breeding and for development of future forests. The practice of seed handling, at all stages from collection to sowing, has an essential influence on the physiological quality of the seed. In floral domain, sexual, i.e. through seed, mode is preponderant. Seeds have been classified in three categories:

- Orthodox :- These retain their viability for long period of storage. This group includes all the seeds which dessicate naturally on the mother plant. The seeds can be dried to lower moisture content (less than 10%) without any serious deleterious effect. The lower the seed moisture content and storage temperature (within limits), longer the survival.
- Recalcitrant :- These lose viability within a very short span. This group of seeds die if these are dried below certain moisture limits (12 to 30%). These are killed when exposed to low temperature (less than 16°C). Even at optimal moist conditions, survival of seeds from this group is limited from

few days to a few weeks. If they are not collected at maturity and sowed immediately, these loose viability.

- Intermediate :- These have shelf life in between the first two categories. These can be dried to seed moisture levels almost similar to that of orthodox seeds without their viability being affected, but can be stored for a limited period under conditions used for orthodox seeds.

A majority of tropical forest species, especially timber trees, produce recalcitrant seeds. The importance and difficulty of conserving recalcitrant seeded species are well recognized. In recent years, awareness over loss of plant diversity has captured worldwide attention. Though several conservation activities of plant genetic resources have been performed, they are very labour intensive and complex. These encompass various tasks such as germplasm collection, storage and evaluation. Germplasm conservation has become necessary to prevent genetic erosion. Major reason for conservation is the need for providing a continuous supply of seed and seedlings for various planting programmes. The current trend the world over is to establish forest plantations in order to reduce the logging impact on natural forests with greater emphasis towards the conservation of existing biodiversity. One of the major constraints with tropical forest management is the availability of good quality seeds. For establishment of successful forest plantations, there is need for quality planting material that are selected from elite mother trees in order to produce seedlings that have higher survival capacity and give greater economic returns. Presently, whilst recalcitrant species can be conserved *in situ* in national parks and forest reserves, some have been taken out of their natural habitats and conserved *ex situ* in arboreta and botanic gardens. In the field of biotechnology, many other complementary methods for *ex situ* conservation have been explored, including cryopreservation. The success of biotechnological approaches is dependent on following factors:

- The ability of germplasm to survive storage treatment like dehydration, dessication, low and ultra low temperature stress etc.
- *In vitro* morphogenetic competence.
- The development of widely applicable, routine and economically viable conservation methods.

For conservation purposes, recalcitrant seeds should be collected from healthy trees with good shape and form. Several techniques of seed collection, namely ground collection, shaking seed bearing branches, free climbing or climbing the trees using equipment can be done. Seeds meant for conservation are best collected freshly from the tree crown by climbing. Criteria for collection of good seeds should include mature seeds, uniform in colour and size and health. Most recalcitrant seeds respire intensively because of their high moisture content and hence require good ventilation. If large quantities are closely packed, suffocation, physiological breakdown, fungal growth and overheating will take place destroying the seed. On the other hand, seed will also deteriorate rapidly if moisture content is reduced too much or too rapidly. This is likely to happen during transportation in open vehicles due to air movement. Plastic bags can be used with top open or with perforated sides. Loosely woven jute bags can also be used. Temperature below 16ºC and above 32º C should be avoided. Shading from direct sun is recommended. Collection journey should not be more than three days. Any germinated seed during collection should be separated from the batch and can be planted right away in nursery and can be later used as a source of shoot tips for cryopreservation. It is essential that the batch is kept moist during transport. A mobile seed- seedling chamber has also been devised for transporting recalcitrant seeds. The chamber is equipped with sensors that control the chamber environment including temperature (20+ or - 5ºC), humidity (80 to 95%) and light as needed. As much quantity of seed should be stored as available because viability is bound to go down. For storage, the strategy is to desiccate recalcitrant seeds to just above critical moisture content, treat with effective fungicide and then store in semi-sealed packaging so that adequate gaseous exchange takes place but moisture loss is restricted. With seed chamber, seed viability in a slow growth condition can be maintained for periods longer than a year. Following techniques are also being advocated.

- Seedling chamber conservation :- Freshly collected seeds are surface treated with a 0.1% Benlate/Thiram fungicide and allowed to germinate under ambient conditions in containers kept moist using wet tissue papers. Once the radicle has emerged, these are inspected for insect attack or fungal infection. Infection-free germinated seeds are loosely packed in polythene bags or plastic boxes containing sterilized sand or lined moist tissue paper and stored in a specially constructed seedling chamber. This chamber has a temperature, humidity and light control and alarm system. The temperature is maintained at 16ºC, the relative humidity at 80% and the photoperiod maintained for four hours each day. The germinated seeds placed in these chambers develop slowly and barely attain the height of 20-25 cm. To date, 20 recalcitrant seeded species have been tested and the storage period obtained was between four and 24 months as reported by Krishnapillay and Marzalina. When these seedlings are removed and planted in polybags, these require weaning in at least 70% shade for a period of 2-3 weeks before these can be placed in direct sunlight. Post storage survival percentages are between 60 to 80% when these are planted out. This method of storage is advantageous when there is an excessive seed fall and can also be used to keep germplasm during long periods of unfavourable weather and delayed replanting programmes.

- Storage of seedlings on forest floor:- This method is not new and has been practiced by nature since times immemorial. In this technique, areas of forest floor are cleared and freshly collected seeds are sown in these areas. These seeds germinate and develop into seedlings. Due to subdued light conditions, the seedlings develop very slowly and can remain within manageable heights for a few years. *Hopea odorata* has been found to grow to less than 10 cm in height in five years. Once these seedlings are potted into polybags and placed under nursery conditions, these rapidly start developing and begin to increase in height. This approach could be adopted to keep stock of planting materials with minimal cost and space. Protection against field pests and diseases and improvement of transplanting techniques is desirable to exploit the potential of this technique.

- Cryopreservation:- Orthodox or semi-recalcitrant seeds of tropical species can be cryopreserved without much problem as their lowest safe moisture content can be reduced to less than 20%. However, when truly recalcitrant seeds are dried and directly plunged into liquid nitrogen, there is no recovery. Many tropical recalcitrant seeds are large in size with varying moisture content and often attached with appendages for dispersal. These seeds do not tolerate a reduction in moisture below a certain level without loss of viability. More research is needed, but exised embryos can be cryopreserved. Controlled freezing, cryoprotective dehydration, vitrification and encapsulation-dehydration are some of the techniques. Embryos are dissected aseptically and left to partially dessicate in laminar flow for 4 to 6 hours. More precise dessication is possible by placing the embryo in a stream of compressed air or in an airtight container in silica gel. The dessicated embryos are placed in cryovials for storage. Slow cooling methods give good results which requires lowering of temperature at the rate of -1°C per minute to - 40°C followed by direct plunging into liquid nitrogen. Slow thawing in a laminar flow cabinet has a beneficial effect on survival. Cryopreservation of tropical tree shoot tips or apical meristems involves using alginate encapsulation . Vitrification is another technique. It involves placing the samples in extremely concentrated cryoprotective solutions and ultra-rapidly freezing these. The intracellular solutes vitrify and form an amorphous glassy structure. Somatic embryos can also be cryopreserved.

Tissue culture is a form of vegetative propagation. Whether it will be most rewarding to propagate a plant by seed, macro propagation or tissue culture will often not only depend on the plant species, but also on the development of proven techniques, relative cost and agronomic objectives.

Tissue culture is emerging as a very potent tool. While the technology is well established in agriculture, floriculture and horticulture, foresters are still skeptical. The scenario is slowly changing and tissue culture technology is making inroads in forestry sector. The process of de novo organ formation is called organogenesis or morphogenesis (development of form). Organised growth contributes towards the creation or maintenance of a defined structure. Unorganised growth is seldom found in nature, but occurs fairly frequently when pieces of whole plants are cultured in vitro. Tissue culture is commonly used as a collective term to describe all kinds of in vitro plant cultures. Complete new plants can be derived from tissue cultures in three ways.

- From pre-existing shoot buds (meristems).

- Following shoot morphogenesis when new shoots are induced to form directly from explanted tissues of the mother plant.
- Through the formation of stomatic embryos which resemble the seed embryos of intact plants.

Some of the advantages of micro propagation are

- Propagules can be free from infections.
- Clones can be produced on a mass scale.
- Production can be all the year round.
- Propagules can be stored for longer period.
- Transportation is easy.

Some of the disadvantages are

- Advanced skills are required.
- Plantlets sometimes have undesirable characteristics.
- The chances of producing genetically aberrant plants may be increased.

Although the new genotypes which can appear as a result of tissue culture are largely unwanted, there are occasional beneficial variants amongst cell lines or regenerated plants which can be isolated by selection. The extent of undirected genetic change can be increased by appropriate cultural techniques, or by the administration of chemical or physical mutagenic agents. In contrast to the largely controlled variation in tissue cultures, genetic manipulation techniques developed over recent years, have begun to allow directed and desired changes to plant genotypes.

Factors influencing the growth and morphogenesis of tissue cultures can be grouped under three main headings.

- Genotype.

- Environment
- Tissue dependent factors.

There are numerous examples in literature in which the results obtained during tissue culture plant have varied from one variety of plant to another.

Composition of medium, growth regulators, gaseous environment, temperature, humidity, light etc. affect the process of tissue culture. Plants are dependent on light for :

- Photosynthesis.
- Photomorphogenesis.
- Phototropism – i.e. the growth response of plant which is induced by light.

The three qualities of light which most clearly influence in-vitro growth and morphogenesis are:

- Wave length
- Flux density
- Photo period

For example

- Blue light inhibits axillary shoot proliferation
- UV light can induce specific enzyme activity
- Yellow-red light induces shoot proliferation

Tissue dependent factors can be grouped as:

- The part of the plant's total genetic code that is expressed (senile or juvenile tissues)
- The manner in which mother plants have been treated.
- The physical and physiological condition of the explants used.

In vitro culture can be used for inducing rejuvenation.

Realising the potential of tissue culture, National Afforestation and Environmental Board (NAEB) and Department of Biotechnology (DBT), Government of India constituted a task force to explore the possibilities of this technology for forestry sector in India. As per the recommendations of this task force, mandate for development of protocols for different tree species has been given to different research organizations. Now, a number of private players are in the field and it will be advantageous to outsource the technological aspect as well as supply of plants of tissue culture origin.

Genetic modification (GM) is one futuristic technology which cannot be ignored. Non-food crops play an essential role as fuel, fodder, building material and countless domestic and industrial processes that are derived from products of vegetable origin, both cultivated and wild. Present and future demand for raw material has been an important factor in determining research agendas and investment strategies within the agricultural and forestry sectors. With non-food crops, GM aims predominantly at the production of plant material for large monocropping systems. Much public concern has focused on the long term effect that GM food might have on human health. Impact of non-food crops have largely been lost sight of. Even though they may not be consumed directly as food, some still find their way into the food chain. Whether eaten or not, transgenic non-food crops have environmental and socioeconomic impacts just as food crops. However, most regulations focus on food crops and products. So there are few labeling systems in place that would enable consumers to choose between GM and non-GM wood. The growing consumption of paper and timber has given an impetus for development of GM trees tailored to the needs of plantation forestry. The long-term effects of GM trees and other non-food crops are difficult to assess. Trees are particularly difficult

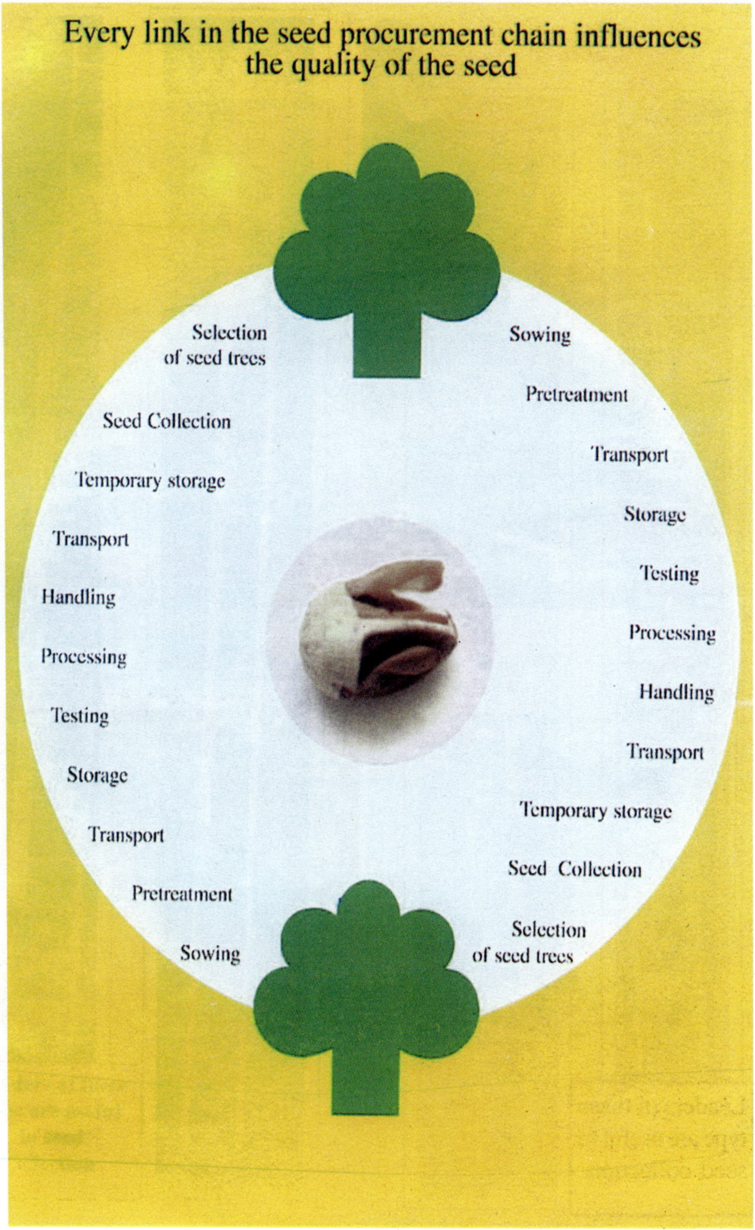
Every link in the seed procurement chain influences the quality of the seed
Selection of seed trees
Seed Collection
Temporary storage
Transport
Handling
Processing
Testing
Storage
Transport
Pretreatment
Sowing
Sowing
Pretreatment
Transport
Storage
Testing
Processing
Handling
Transport
Temporary storage
Seed Collection
Selection of seed trees

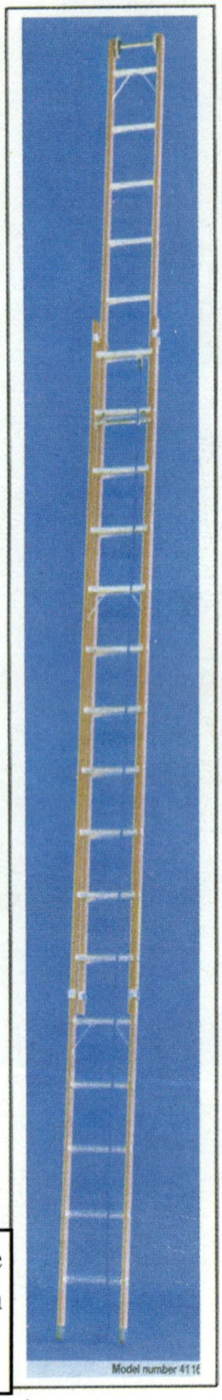

Leaders of these type are useful in seed collection

Seed Grader

Seed Germinator

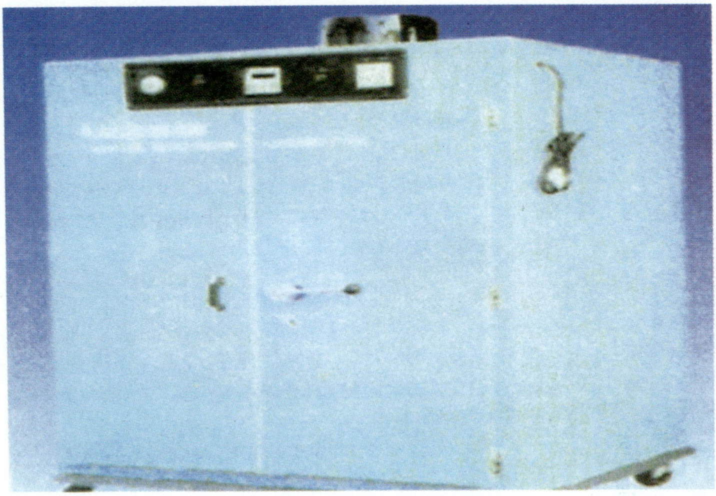

Seed Driver

Computerized Seed Counter

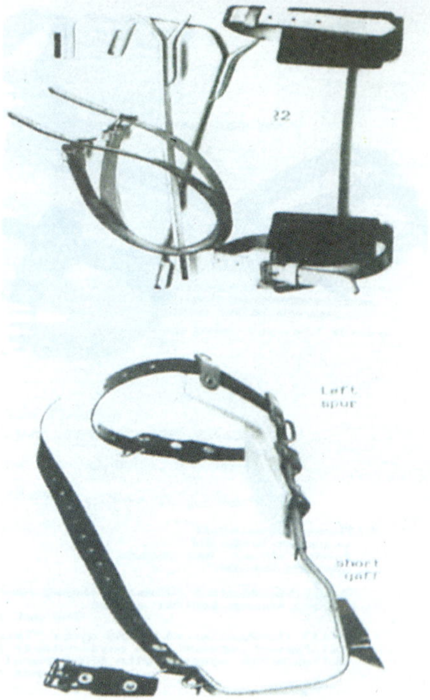

Tree climbing equipment for seed collection

Intelligent Seed Moisture Meter

to monitor. Often grown in isolated areas, they live much longer than annual crops and unwanted traits may appear after several generations. Preventing cross contamination with wild species and trees have ingenious ways of propagating themselves over long distances. Sterility technology has yet to be perfected and some doubt whether it is wise to try. Trees and soils have become a focus of attention in the climate negotiations for their capacity to store carbon and hence slow down the accumulation of greenhouse gases in the atmosphere. If trees and soils are included in the Climate Convention as carbon sinks, it will certainly promote GM plants like faster growing trees or herbicide resistant crops used in conservation tillage. One area of concern is the difficulty in obtaining comparative data on GM crops. Information on the research being carried out, the extent GM varieties have been introduced into the environment, and the way they are being monitored, have not been systematized and are, therefore, fragmented. This poses a huge challenge for effective implementation of the Cartagena Protocol on Biosafety. Given expansive scale of forests and plantations and the critical role that forests play in fostering biodiversity and regulating climate, the discussion about GM trees should include the impact of transgenic plantations on natural forests. The complex of symbiotic dynamics between species is a mystery, which makes rigorous risk assessment all but impossible and complicates any study of the effects of GMOs on wild forest biodiversity. Field trials are underway in countries like Chile, Indonesia, South Africa, New Zealand and China for designer trees and the focus is on Poplars, Radiata Pine and Eucalyptus. ArborGen and Tree Genetic Engineering Research Cooperative (TGERC) are some of the giants engaged in research into GM trees. Their philosophy is that biotechnology clearly offers the potential to improve the sustainability of the earth for a population soon to be ten billion plus. FAO has projected that if consumption of forest products does not change, there will have to be an annual increase of between 1.1% for pulp and 2.4% for paper and paperboard to meet current market trends. Protagonists of GM trees argue that increasing population pressure poses a threat to the last remaining ancient forests which can only be alleviated if wood supplies were met through GM plantation forestry and producing GM trees tailored to the consumer will have environmentally friendly spin-offs.

8 Go For Multi Tier Planting

With bourgeoning population, land holdings per capita are shrinking. Available forest area is also limited. In urban areas, the simple solution for limited availability of land is to go for high rise buildings with basements. Same approach is advisable in forestry and agroforestry. Light demanders with tall boles can be grown so as to occupy top canopy. Shade bearers can be grown as middle canopy. Herbs and grasses can form the ground flora. A large number of models have been developed by research organizations and agricultural universities. A suitable model can be adopted for specific agro climatic zone. Agroforestry is the collective name for land-use systems where deliberate use of woody perennial plants and other crops and animals is done on the same area of land in some temporal and spatial arrangement where there is both ecological and economic interactions between trees and other components of the system. Agroforestry systems with their element of intercropping are in essence ecological models similar to natural systems and have, therefore, an element of sustainability associated with them. Farmers have been practicing some form of agroforestry since times immemorial. Agroforestry represents the integration of the best of what we know, the knowledge and skills of forestry and agriculture. Skills which have, in large part, been managed separately, compartmentalized by disciplines and by institutions. With agroforestry, foresters, agronomists, biologists and many other specialists have the opportunity, through a voluntary non-regulatory approach to share with land owners a resource management approach that is logical, science based, and user friendly. At the same time, it must be highly customer driven. Agroforestry is relevant in multiple scales – in space and time. Agroforestry is "working trees" – i.e. planting the right tree in the right place for a specific purpose. Agroforestry can be an important tool for sustainability – sustainable forestry, sustainable agriculture and sustainable development. Agroforestry is partnership friendly. It crosses boundaries, blends, harmonizes and maximizes resource outputs and amenities. It is a win-win activity. Fish and wildlife habitats are improved through agroforestry. Riparian habitats are crucial to many wildlife species. Agroforestry can provide systems that provide economic incentives for developing and improving the vegetation on these areas. Aquatic and wetland habitats may seem unrelated to agroforestry programmes. Water quality and quantity in a watershed is, however, closely related to how the agricultural land in that area is managed. For example, tree and shrub plantings along filter strips can greatly enhance water quality downstream.There are four key principles in agroforestry:

- Intensive land use
- Integrated systems
- Biological interactions
- Optimization of benefits

Agrisilviculture, hedge rows, homestead forestry, biofence systems, coffee-based agroforestry, cashew based agroforestry, eucalyptus-based agroforestry, mango-based agroforestry, bund plantation, coconut-based agroforestry, palmyra palm based horti-agriculture are some of the popular traditional models. Some single species (like Santalum album, Acacia mangium, Gmelina arborea, Tectona grandis) dominated models are also being adopted. Change is the law of nature and new models are replacing old ones. In Vishakhapatnam district, oil palm plantations are taken up with field crops of paddy. Teak and casurina are planted with chilly, sugarcane, banana etc. Mango is planted with groundnut and mesta (Hibiscus cannabinus) in initial years. Curry leaf is grown along with teak, mango and other field crops. Multistoried or mixed plantations of cashew plus mango, Acacia mangium plus banana plus pineapple and cashew plus teak plus pineapple can also be seen in the fields of some enterprising farmers. In Koraput, Bastar and some other districts of Andhra Pradesh, in dry lands mango is being planted at 8 metre spacing and space in between is often used for planting straight and fast growing trees like Tectona grandis, Acacia auriculiformis, Acacia mangium, Leucaena leucocephala, Casurina equisitifolia etc. For biofence, Agave sisalana, A. Americana, Acacia concinna, Bamboo, Duranta ripens, Erythrina indica, Euphorbia spp. Glircidia, Ipomia, Jatropha and Pithocellobium dulce are the common species. Most of the tree species under traditional practices have strong socio-cultural attachment with tribal livelihood besides their ecological and economic potential. Mango, *Mahua*, Jackfruit, *Karanj* and fish tail palm are retained beyond their economic rotation and reflect the priorities of tribals.

Farmers in parts of UP, Uttaranchal, Haryana and Punjab have enthusiastically adopted Poplar as an agroforestry crop. One suggested model is Poplar with sugarcane and mentha. In this intercropping system, sugarcane is cultivated in second week of January. Then in third week of January, ETPs of Poplar are planted. Thereafter, mentha is cultivated after two to three weeks. At the spacing of 5mX4m, 500 ETPs of Poplar per hectare are required. Requirement of sugarcane seed per hectare is 75 quintals. Three quintal of mentha suckers per hectare is needed. The land is ploughed and disc harrowing is resorted to followed by cultivator and leveler operating. Irrigation channels are made 5m apart (centre to centre). Sugarcane is cultivated at an interval of 90cmX90cm. Mentha is sown in rows at a distance of 45cm. Healthy suckers of Mentha are cut into 8-10cm long pieces and put about 5-6cm deep in soil and well covered. Plantation of Poplar is carried out in pits at a spacing of 5mX4m. Frequent but light irrigation is recommended. Two to three weedings should be performed before first harvesting of Mentha. First cut of Mentha can be taken 100 to 120 days after sowing. Oil production of Mentha is highly influenced with the time and the stage of cutting. It is better to cut crop on a sunny day. After this at least one hoeing is must so that the roots are well mixed with soil. Sugarcane can be harvested as it becomes mature. Poplar is ready after 6 to 7 years. Recommended varieties of Poplar are S7C4, S7C8, S7C15, L-34, L-49, G-48, Udai, Bahar and Kranti. Mentha arvensis (CIMAP MAS – 1), M. pipereta (Kukrail), M. spicata (CIMAP –MSS -5, Urca and Nira), M.citrata (Kiran) are preferred. For sugarcane Co-88250, Co-91269 and Co-Pant 90223 are recommended.

The essential oil-bearing crops viz. Lemongrass, Citronella java and Palmarosa have traditionally been grown as a monocrop. Now, the trend is to integrate these with fast growing trees like Poplar using agroforestry system. Underplanting of aromatic grasses with Poplar have yielded following economic returns.

- Poplar alone - Rs 34,450 per ha. per year.
- Lemongrass alone - Rs 18,350 per ha. per year.
- Citronella alone - Rs 22,500 per ha. per year..
- Palmarosa alone - Rs 20,805 per ha. per year.
- Poplar with Lemongrass - Rs 50,115 per ha. per year.
- Poplar with Citronella - Rs 40,960 per ha. per year.
- Poplar with Palmarosa - Rs 43,640 per ha. per year.

Saline, Sodic and Saline-Sodic soils are commonly termed as *usar* which are highly degraded. The extent of *usar* land in India is estimated to be 71,64,000 hectares. Extensive research has been done in India and Israel to improve the productivity of such type of land and results have been very encouraging. Narendradev

Agricultural University, Kumarganj, Faizabad, U.P. has developed a model for raising orchards of *Aonla* (Emblica officinalis). It is based on multiple land use pattern and the results are for all to see at site. The varieties recommended are Narendra – 6, Narendra - 10, Narendra - 7 and Chakaiya. Other fruit plants that can be grown in *usar* areas are Psidium guava (variety Lucknow -49) and Zyzyphus maurtiana (variety Gola, Banarasi karaka, Kaithali, Umran), Aegle marmelos (variety Narendra bel -4, Narendra bel - 5and Narendra bel - 9), Crissa carandas and Grevea sabiniculis. The problem may be in marketing and storage because the keeping quality of the fruits is poor. The solution is very simple. Foodgrade polythene bags with tongue and groove arrangement at the open end may be used. The bags are lined with crystals of bioceramic which emit far infrared (FIR) rays. The bags are airtight and FIR rays have a property of preventing proliferation of bacteria, fungus and virus and it is these organisms that cause rotting. FIR rays are the heat producing part of sunrays and help in ripening of partially ripe fruits. Thus, partially ripe fruits can be plucked and stored in these type of bags to get deliciously ripe fruits after 10-12 days at a very economical cost as the bags can be recycled till tattered due to mishandling.

While collectively planting several species at one spot alleleopathic considerations should be kept in mind. This science has not been given the consideration it deserves. In nature different species exude different chemicals. These chemicals are compatible for some species while antipathic for some others. Nature has devised this mechanism to eliminate competition and promote symbiotic relations. In field tested modules of multitier planting, this aspect gets sorted out. But for models in conceptual stage, statistically designed field testing is must. Some recommended agroforestry models are being listed in the table below:

Topography	Agroforestry System	Components						Recommending Organisation
		Tree		Crops		Pasture		
		Forest	Fruit	Commercial	Others	Grass	Legume	
Hill slopes of eastern ghats	Agrihorti silviculture	Sissoo	Guava		Soya-bean alternating with wheat			RRTTS, Semiliguda
All tracts	-do-	-do-	-do-	Ginger/ Turmeric				-do-
Hill slopes of eastern ghats	Silvipastoral	Cassia siamia			Rice, Bean and cowpea	Dina-nath		-do-
Sloping Land	Cashew			Cinnamon, tapioca, coffee, cardamom, turmeric, ginger, sisal, black pepper on cashew				CS & WCR & TI Research Centre, Koraput
Sloping land	Teak	Mango, Guava, Banana		Coffee				CARS, (IGAU), Jagdalpur
Old orchards	Mango			*Arhar*				NDA&T Kumargunj
New orchards	*Aonla*			*Haldi, Barseem*				-do-
-do-	*Ber*			*Jimikand,Barseem*				-do-
-do-	All fruits			*Adrakh*				-do-
-d-	*Ber*,Guava			Onion				-do-
-do-	*All fruits*			Garlic				-do-

- List of various MPTS suitable for different regions.

A. Himalayan Region

a. Kashmir

1. Almond/apple in rows with saffron/vegetables.
2. Walnut (on drains and bunds) with maize and beans.
3. *Robinia pseudoacacialCeltis/Ailanthus/Morus/ Salixipoplar* with wheat and paddy.
4. Leucaena/mulberry/poplar with paddy-wheat.
5. *Acacia nilotica, Eucalyptus* with pulses-maize/fallow/citrus.

B. Himachal Pradesh and Uttar Pradesh

1. Mixed plantations of fodder, fuel, timber and forest trees with vegetable crops under irrigation; *Eleusine coracana* and wheat in drylands.
2. *Kydia calycinalMorus* with paddy-wheat on terraces.
3. *Sapium insigne* with paddy-wheat.
4. *"Boehme ria rugulosa, Grewia optiva, Terminalia, Celtis caucasica and Parkia roxburghii* with paddy-wheat or maize-wheat or ragi/jhingora in terraces.

C. North-Eastern States

1. Arecanutlcitrus with black pepper, banana and papaya.
2. Arecanut with pineapple, betel vine/citrus with cardamom.
3. *Grevillea robusta* with paddy.
4. Bamboo, *Parkia roxburghii, Mangifera indica, Syzygium cumini, Emofficinalis and Morus alba* with vegetables.
5. *Dipterocarpus macrocarpus* with coffee, cocoa and *Terminalia microcarpa.*
6. *Shorea robustalGrevillea robustalAlnus nepalensis* with tea, cardamom, turmeric and ginger.

7. *Indigofera tasmianii* with *Albizia lebbeck, A. procera, A. odoratissima, A. stipulata, A. mollucana* and *A. sumatrama* with tea.

8. *Terminalia mircrocarpa, Bischoffiajavanica* and *Alnus nepalensis* with cardamom.

D. Alluvial Plains and North Eastern Areas

1. Eucalyptus with soybean/maize followed by wheat
2. Eucalyptus with lahi followed by lentils/gram.
3. Poplar with soybean/maize/lahi followed by wheat/lentil.
4. Sugarcane with poplar up to three years followed by paddy/potato.
5. Poplar/Eucalyptus with sugarcane for three years followed by paddy/potato.
6. Leucaena/Sesbania/Poplar with Brassica/Cymbopogon.
7. Dalbergia sissoo and Acacia nilotica with sugarcane for three years followed by wheat.
8. *Acacia nilotica/Tamarix articulata/Agave sisalana/Cajanus cajan* on bunds with paddy and wheat or maize and wheat/Eucalyptus/Poplars/mango/shisham/ber/mulberry.
9. Leucaena/Prosopis *cineraria/Azadirachta indica /Albizia lebbeck/Toona cillata* in mixture with crops.
10. Along the orchards a local variety of Eugenia is grown as a shelter crop to protect the orchard from the scorching sun. The jamoa crop is thinned for fuel and small timber.
11. *Leucaena leucocephala, Shorea Robusta, Tectona grandis,* Champa, Cryptomeria and Chukrasia are the main tree species grown with ginger, turmeric, bamboo, black pepper in vacant places in West Bengal.
12. Shisham and bamboo are grown on the bunds in Bihar. In

early stages, maize is grown and later pulses. Intercropping is confined to better sites where wheat is the principal crop.

13. Eucalyptus with marigold flowers/bajra/castor/vegetable/ sugarcane in Gujarat.
14. *Casuarina equisetifolia* on field bunds with cabbage and mango in coastal areas.

E. Arid and Semi-Arid Region

1. *Faidherbia albida* with cereal and grasses.
2. *Albizia lebbeck* with moth bean/groundnut, sesamum/castor/ cluster bean.
3. *Azadirachta indica* with moth bean/cluster bean/groundnut/ sesamum/castor/pasture grass or forage legume.
4. *Prosopis cineraria* with pearl millet / green gram / sesamum / castor / pasture grass.
5. *Zizyphus* mixed with moth bean and pearl millet.

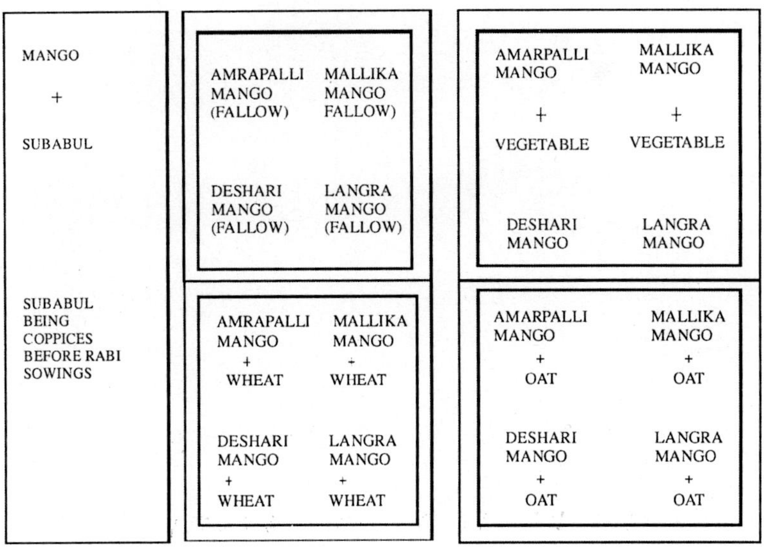

This hand held pit digger can be handy in planting operations. Deep soil working helps the roots in establishing in refractory sites

Vetiver is good soil binder and is of economic importance

Kannauj in Uttar Pradesh has been an important centre for scents since times immemorial. Fragrance and Flavour Development Centre at Kannauj has done a yeoman's service in doing research and disseminating knowledge in this field. It is rather sad that as the biggest landholder, forest department has not taken advantage of the knowledge base to improve the productivity of so called waste lands or, more appropriately, wasted land. Often forester take shelter behind a clause of Forest Conservation Act 1980, that cultivation of aromatic and medicinal herbs imply non-forestry use of forest land. It will be in the interest of nation that such dogmas are shed and acts are interpreted in spirit rather than the letter. FFDC has done some work on Mentha which can be grown with potato. Yield is 100 to 150 kg/ha depending on species. The oil is used for remedies of cough, balms and pain relieving formulations. Palma Rosa is another aromatic perennial grass which yields ginger grass oil. The cost of cultivation is Rs18,000 to Rs 20,000 per hectare and annual income is Rs 50,000 to Rs 60,000 per hectare. Citronella Java is also a perennial grass yielding Citronella oil. Cost of cultivation is Rs15,000 and income is Rs 45,000 to Rs 50,000 per hectare. Indian Basil or *Tulsi* yields oil, fragrants and medicines. There are more than 60 varieties of *Tulsi* out of which following six are common in India:

- Ocimum basilicum or *Mithi Tulsi*
- O. americanum or *Kala Tulsi*
- O. garetisimum or *Ram Tulsi*
- O. climenchericum or *Kapoor Tulsi*
- O. sanctum or *Gharelu Tulsi*
- O. viridi or *Jangli Tulsi*

First two are commercial. Cost of cultivation is Rs3500 and income is Rs 13,000 to Rs 14,000 per hectare per annum. Lemon grass is a perennial yielding Citrol. Cost of cultivation is Rs 12,000 to Rs 15,000 and income is Rs 40,000 per hectare per annum. Vetiver or *Khas* yields Vetiver oil which is used in perfume, *pan masala,* cold drink, cosmetic, indigenous medicines and tobacco industries. Net return per hectare per annum is about Rs 50,000. This institute offers following services:

- Supply of planting material/seed.
- Production/processing facilities on job work/loan basis.
- Creation of fragrance and flavour for various end uses.
- Quality assurance and analytical services.
- Training in various disciplines of essential oil, fragrance and flavour industry.
- Field survey and feasibility studies.
- Consultancy services.
- Application of fragrance and flavour in end products.
- Marketing of essential oils, its fractions, fragrance and flavour.

The organization proposes to add following new dimensions to its activities:

1. Establishing demonstration farms in different agroclimatic zones of the country for various aroma-bearing crops.
2. Opening of quality assurance and technology diffusion centres in other areas of concentration of aroma-bearing crops.
3. Establishing scarce raw materials bank for small-scale industry.
4. Establishing a marketing division to assist new entrepreneurs/ farmers.
5. Taking up development work for aromatherapy and flavours of Indian origin.

6. Standardization of Indian *attars* and adoption of ISO systems for manufacturing and quality assurance.

Beekeeping along with agroforestry is another interesting option. It can be lucrative too. For instance, one farmer on the outskirts of Ludhiana has 20 acres of land. During 2002 he exported 1500 tonnes of honey earning Rs 4 crores. This enterprising farmer buys honey from small-scale operators as well and encourages contract farming. He supplies equipment, bees and training with a buy-back arrangement. As per this farmer, if one invests Rs 2 lakhs in honey production, within two years the investment can be recovered. Of course, with this entrepreneur, quality control measures are stringent. To identify export markets, this entrepreneur travels extensively and takes part in six international honey-related exhibitions. He avers that in honey, there is more competition amongst buyers than among sellers and is hard put to honour export obligations. Total production of honey in India is to the tune of 15,000 tonnes. Nearly 12,000 tonnes is consumed locally and 3000 tonnes is exported. There are only two serious honey exports in India. Contrast this scenario with figure for China and Argentina each producing 80,000 tonnes. Value added products like squash, pickle, herbal honey, orange honey, lychee honey, lime honey, saffron honey etc. enhance profitability and marketability. Khadi Commission gives 30% subsidy for honey-related business projects. Indian Farm Forestry Development Coop. Ltd. (IFFDC) operating in four districts of UP are doing good work and are worth emulating.

9 Forest Fires Are Manageable

As per the data of Forest Survey of India for 1995, 53.54% of forest area is annually affected by incidence of forest fire. Ecopyrology is the science that deals with starting forest fires states that eliminating fire from ecosystems is like shutting out the wind and the rain. Fire suppression alters ecosystems. Open habitats favoured by wildlife get rarer, and there is a risk of loosing the biodiversity it sustains. Ecopyrology propagates the use of fire in land management and controlled burning is a forest management practice. Many forests are born of fire. Fire is sometimes used as a tool for pastureland improvement. Under certain circumstances, quality of forage is improved, flowering and seeding. Collection of flowers, tender leaves, honey, lichens and seeds is facilitated. Certain plant species have adapted themselves to cope with and even benefit from wildfire. Trees like Chir pine now inhabit areas where fire means survival and not extinction. Process of regeneration is accelerated because favourable growing conditions are induced. By removing litter, landscape can be beautified. Removal of litter accumulation due to fire can mean removal of physical hindrance for establishment of tender seedlings. Organic debris can be converted to nutrients in the shape of ash as a result of fire. Thus nutrients are recycled resulting in a spurt in growth of plants. Often weeds get destroyed in fires. Many useful plants flourish in full sunlight. Herbivores prefer the fresh flush of succulent shoots that sprout after fire and weight gains are significant vis-à-vis in unburnt areas. Woodpeckers population is significantly enhanced in burnt areas because bark beetles and other insects colonise newly-burnt trees and woodpeckers come to feast on these. Seeds found after fire prove useful for many mammals and birds. These seeds are leftovers that did not germinate because these fell on a shaded or crowded forest floor. Many seeds – particularly grasses - with hard outer cover get singed in fire which facilitates germination. Dissemination of seeds from mature cones is also facilitated by fire because the cones open only when heated. Many species are better able to resist singing by fire due to thick bark which insulates the inner tissues from fire. In some conifers, the needles are also fire resistant. Thus, fire is an important and powerful tool of habitat management in a way. But unplanned and forest fire beyond control produce undesired results. In order to obviate such an eventuality, possibility of wildfire has to be warded off. When the Britishers introduced forest management the tools adopted for fire conservancy were:

- Control burning before summer
- Clearing of fire lines to contain conflagrations
- Counterfiring
- Controlling wildfires by primitive tools with the help of local people. This included beating of ground fire with brooms fashioned out of local bushes.

The system was more or less effective so long as grass root level functionaries were sincere and remained at their posts during fire season. Due to reasons stated elsewhere, sincerity of forest officials got eroded, absenteeism from place of posting became rampant and the bond of kinship between local population and forest officials weakened. False bills of fire line clearance and controlled burning became the order of the day. Net result was alarming increase in cases of wildfire during summers. Instead of being put off by conscious efforts, by and large wildfires are allowed to die their natural death at the edge of forest or due to summer rains. As usual in India, enquiries are conducted and the matter is allowed to fade from public memory. Excuses are invented to let the guilty off the hook. Forest officials also let the budgeting process drift from

Tractor drawn Containers Can Be Useful In Transporting Slash From Fire Lines And Forest Floor For Briquetting To Make The Operation Economically Viable.

Spreyer Can Be Useful In Firefighting For Application Of Foam. Other Inputs Can Also Be Sprayed.

This leaf sucking device can be useful in keeping fire lines and forest floor clean during fire season.

This light weight portable water pump can be useful in fire fighting and spraying in remote areas.

Collapsible water tanks are useful in fire fighting and spraying in remote areas

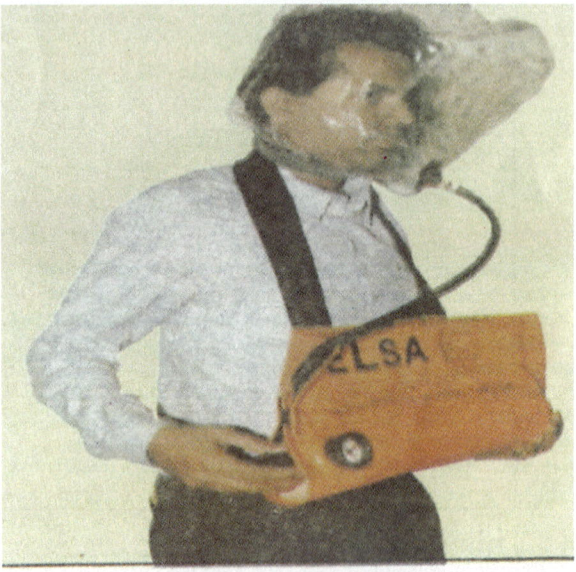

Self supporting life saving systems can be a morale booster

Protective clothing can save lives during forest fighting

It can be useful in minimizing fire hazard and also for spraying foliar fertilizer in plantations

bad to worse in the garb of plan and non-plan expenditure. The department started clutching at straws in the form of Central grants and externally aided schemes. The motive has been garnering funds and fire conservancy objectives were incidental. Take the case UNDP project for UP. Helicopters and planes were inducted but were used invariably for movement of VIPs. Huge inventory of heavy machinery was allowed to pile up which was hardly suitable for the terrain and maintenance facilities. The entire stock can still be seen rotting at Haldwani in Uttranchal. No wonder, the project became moribund once the flow of external funds dried up. Central funds were utilized primarily for payment of wages of seasonal fire watchers wireless systems and propulsion charges of vehicles. Funds earmarked in World Bank Project were utilized for wireless systems, equipment and payment to village committees. There had been little or no long term vision, perspective planning or thought about sustainability. Net result is that with drying of funds the department is back to square one. Incidentally, incidences of wildfire took place during tenure of projects also. The severity of wildfires is dependent more on weather conditions and not on projects. Following steps will surely minimize incidences of wildfire:

- Such administrative steps are taken which ensures presence of staff at the place of posting.
- Old fire lines are religiously kept clean. Some of the handy tools are portrayed in Fig. 9 which can be useful in shrub cutting and sweeping forest floor. Debris from shrub cutting can be used for vermicomposting or electric cogeneration to make the operation cost effective. A weed cutter with appropriate blade can be used to clear all the weed and undergrowth of bushes. The implement costs about Rs 27,000 and one person can cover 1 hectare in one day. Both petrol and electric models are available.
- In high risk areas forest floor can be kept clean with vacuum

cleaners. This can be specially useful in Chir pine forests. The litter or needles collected can be used for manufacture of particle boards or fuel briquettes to make the operation cost effective. High risk areas are adjacent to roads and paths. Areas set aside for regeneration are control burnt during winter. If this prescription is followed in letter and spirit, fire hazard will be minimized due to non availability of inflammable material on forest floor.

- At present, the problem during fire season is non availability of water within forest tract. If the suggestions given in Chapter 3 are followed, there will be enough water in strategically located ponds. Fogging, misting and sprinkling devices suggested in that chapter can be utilized to increase the humidity on forest floor and atmosphere.
- If some or all the above steps are taken, two of the three ingredients of wildfire, viz. dry atmospheric conditions and inflammable material will not be there. As a result of this, the third factor i.e. incendiary source – whether match or lightening or friction between twigs or sparks from burning crop residue in neighbouring fields – will be ineffective in kindling a wildfire.
- All the same surveillance is needed in case of mishap in spite of all the precautions. At present the preferred method is patrolling on foot by seasonally employed fire watchers. In plains, all-terrain geared bicycles may be more effective. Some firewatch towers have been erected in recent past. These are seldom being used because manning the towers require dedication and patience. The main purpose of these towers might have been garnering of funds and utilising the same. Anyway, administrative measures need to be devised to use and maintain existing towers. A more effective way is pilotless flying machines with sensors. If simpler solutions are

preferred, balloons are good enough. Detractors may scoff at the idea, but it should not be beyond the technical competence of a nation which has just embarked on a Rs 3600 million project to put a man on moon.

- Quick communication is must for which wireless network is there. The need is to keep it in working order by a dedicated team.

Above discourse makes it clear that if there is a will it is possible to use fire as a good servant and not as a bad master of foresters. Some of the suggested equipments are illustrated in Figure 9.

10 Do Not Be Dogmatic About Wildlife And Vermin

No doubt, habitat destruction and indiscriminate killing has decimated some wildlife species which were glamorous showpieces. This has resulted in backlash which has affected the balance so essential in nature. Take the case of Neelgai or Vanroj. This quadruped of horse family roams as a herd in agricultural tracts and browses and tramples standing crops. The animal is fairly large and strong and is very aggressive. Non-lethal methods of scaring marauding herds are ineffective. At times, farmers trying to protect their crops have been gored to death or severely injured. Simple solution is shooting. But there is one catch. The appendage of "Gai" in the nomenclature has religious overtones for Hindus. Killing by minority community can lead to communal conflagration. So, there is lot of bassing the buck between forest officials and general administration. Sufferers are poor farmers who helplessly watch their lush crop being destroyed for which they have toiled hard for several months. Cases of farmers committing suicide in despair are also reported. It is height of callousness to let drift a matter which has ramifications on human rights front as well. The simple solution is to conduct propaganda blitzkrieg in affected areas highlighting the fact that Vanroj is related to horse family and not cow family. Such blitzkrieg should be followed by group discussions inviting opinions covering the entire spectrum. This should be followed by consensus building. Given the facts, the consensus is bound to be in favour of shooting the marauding herds. The game will provide some nutrition to some sections of society. If per chance, the consensus is not in favour of shooting, the alternative will be a little expensive. The offending animals will have to be trapped with snares or by using tranquillizing guns. The herds can be tracked by helicopters and the darts can be shot from air at close range. The trapped animals can be translocated in wastelands suitably fenced.

The animals should be sterilized so that over a period of time, population gets reduced to manageable proportions. Chemical repellents can also be sprayed on crops to ward off the vermin. This requires further research. Air Force stations are also being threatened because of *Vanroj* walking into runways. They have no dearth of funds and can finance some research on the issue. The problem has bourgeoned because there are no predators and food is plentiful. Some products being marketed abroad as deer, cattle, elk, moose, rabbit and opossum repellent under the brand name of Plantskydd are being described here. The product is available as soluble powder, premixed liquid and in granular form. Animals avoid plants before they bite and not after. It has been developed and is manufactured in Sweden because in 1991 it was found that commercial plantations were being decimated by deer, moose and rabbits. It affords effective plant protection for more than 6 months as indicated in an article in Scandinavian Journal of Forest Research by Bernquuist and Olander of Swedish University of Agricultural Sciences after field trials between 1991 to 1994. Plantskydd's effectiveness is attributable to its tenacity in sticking to plants, even under severe snow/rainfall conditions and has been found very effective against damage by browsing damage. In USA and Canada, Plantskydd's effectiveness has been confirmed in operational field trials undertaken by Forest Industry Cooperators on millions of coniferous and deciduous trees of various species. Even seedlings can be pretreated at the nursery prior to planting. In USA and Canada Ainsworth Lumber Co. Ltd., Michigan Soil & Water Conservation Districts, British Columbia Ministry of Forests, Canadian Forest Products, Chippewa National Forest, Sylvan Vale Nursery, Weyerhaeuser Canada Ltd. are some of the satisfied users of the product. It is made from specially processed edible animal protein (blood meal), vegetable fat, salt and water. One kilo treats about 350 to 450 seedlings and cost in USA is $33/kg. For forestry operations Diaphram Pump Sprayer of 20 litres manufactured by

Cooper Pegler & Hardi of USA is suitable.

Wolf menace is another burning problem. Wolf is an elusive animal and difficult to track. *Kanjar* community is adept at tracking wolves and their services can be utilized in trapping maneating wolves. The trapped wolves can be put in suitably fenced wasteland tracts along with herbivores. Similarly, panthers sometimes become menace in habitations. These are trapped in very crude cages with the result the animals injure themselves. Trapped panthers are translocated in zoos. By and by, the zoos are getting overcrowded with panthers. A better solution will be to suitably fence large tracts of wastelands and put the trapped panthers in such enclosures along with herbivores. Wild boars are also in the categories of vermin and need to be put in suitably fenced areas. Monkeys in urban areas are becoming a menace and trapped monkeys can be put in enclosures. Of course, the enclosures have to be so designed that monkeys have no chance of getting out. All the vermin trapped can be put in one fenced area in suitable predator – prey ratio. Fisher forest in Etawa is one such suitable area which can be suitably fenced. One method of fending off unwanted wild animals (or for that matter domesticated ones) intruding in habitated locations, can be ultrasound pest repellers. Instruments for rats and birds are available. It should not be too difficult to devise instruments for *Vanroj* and monkeys. Elephants are a different proposition because the ultrasound waves may harm humans as well. As a way out, marauding herds of elephants can be tracked from microlight aircrafts or ultralight helicopters and strong beams of ultrasound waves can be directed at herds to deflect the animals away from crop fields.

Elephant herds have also multiplied manifold and cases of man – animal conflicts are rising. But this problem will be a tough nut to crack because of the size, enormous food requirement and wide ranging habits. In Africa, the problem was solved by capturing the animals through use of tranquilizing guns fired from helicopters.

Unconscious animals are lifted on tractor drawn trailers and then on trucks to be released in new locations. The entire operation was telecast on Discovery channel. In the new location, the railway track and private farms were excluded by electric fencing to avoid man – animal conflict. If one is inclined to shun such high tech solutions, then using animal decoys is an age-old practice to trap wild animals. Using tame elephants to lead wild herds towards traps during *Kheda* operations is a prime example. Recently one programme was telecast on Discovery channel depicting a tame giraffe being used to lead wild herds towards traps for eventual translocation from overpopulated reserves.

Unfortunate part of wildlife management in India is that it is shackled by dogmas and fixed ideas. There is no dearth of solutions if one thinks with an open mind. For calculating cost benefit ratio, the hidden costs involved in procrastination and alternatives should be taken into consideration before taking decision.

Water bodies are important wildlife habitats. These are being destroyed due to silting, eutrofication and growth of vegetation like water hyacinth. First two problems require an integrated management approach. Dredging of silt may be an option but it may destroy many organisms in the process. Careful handling may reduce this risk. Wetlands form the interface between terrestrial and aquatic systems and perform a number of important functions of ecological importance and socio economic relevance. This is precisely the reason why they are threatened. During lean season these become prone to encroachment. There is the ecosystem, functional (flood control, groundwater recharging, effect on microclimate), habitat (for flora and fauna) and anthropogenic (fishing, tourism). Wetlands are fragile ecosystems because of lack of self-cleaning ability. Water hyacinth is a very problematic free floating fresh water plant. It restricts the flow of irrigation water, interrupts the free movement of boats in lakes, interferes with fishing

Even in rugged Himalyas, wildlife is not safe from poachers

Ultrasonic Repellers for Birds & Monkey

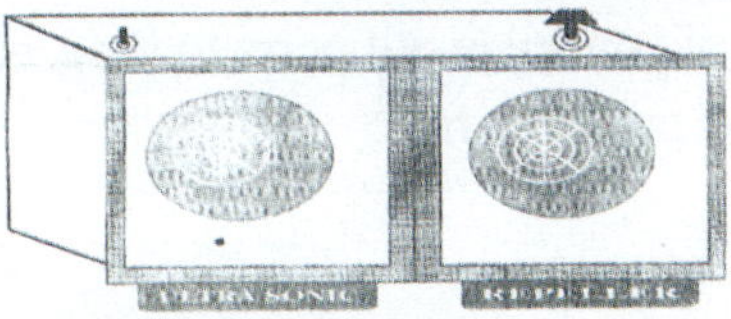

FEATURES : SPECIFICATIONS :
No Chemicals, Nets. Firearms
Easy to use
Automatic Unattended operation
Maintenance Free
Solid State Electronics
220 V. AC Operated
Power Consumption : 5W
Continously Auto Variable
Protect upto 4000 cub.ft. indoor & 2000 cub.ft. out door. Apprx.
Adjustable Immunity check Knob
High Pressure sound, Double Speaker
Environment friendly, Ecologically safe

USES AT :-
Ware Houses
Hangers
Loading Docks
Arenas
Indoor Stadiums
Building Lawns
Granaries
Gardens
Grapes yards
Fruit farms apple, pear, pitch, orange etc.

Bigger models can be useful against blue bull, boars and elephants.

and pisciculture, degrades quality of water and induces a tendency in water bodies to dry because of increased silting. Application of chemicals or mechanical control measures are feasible but cost of repeated application is high. Biological control is permanent, self sustaining and regulates the weed growth by natural predation. The natural enemies of this weed are Neochetina eichhorniae and N. bruchi. Though this method is slow in action, it is long lasting, cheap and ecofriendly. The adults of both these weevils feed on the leaves of hyacinth and deposit their eggs below the epidermis of petiole and laminae. N. bruchi prefers basal ligules of outer leaves while N. eichhorniae prefer small tender leaves. The whitish eggs of these species hatch within seven days. The grubs are cream coloured with yellow orange head. They tunnel into the petiole crown and feed extensively. Some pathogens may also enter through these tunnels and may weaken the plant. During pupation, they move to the live roots and pupate by cutting a lateral rootlet for making a spherical cocoon around themselves. The adults of N. bruchi and N. eichhorniae live for 134 and 142 days and lay about 680 and 980 eggs respectively during their life cycles. The weevils can be reared under controlled condition on the weeds and can be used to control it. This method of control is self perpetuating, ecofriendly, environmentally safe and pollution free.

11 Holy Cow And Other Quadrupeds

It is estimated that animal husbandry is contributing 10% of national income by way of milk, eggs, meat, hides, wool etc. Several practices of production are in vogue and are well integrated in day to day life of farmers. For making the production system a credible enterprise, one has to follow certain guidelines listed as under.

- Good number of physical resources are needed to meet demand and supply
- Geographical location which includes housing
- Size of the unit plays a good role by way of managerial practices
- The role of agro-climatic conditions is important specially in case of dairy industry
- Inputs should be economical
- The enterprise should be planned as per the need of the consumer
- Marketing facilities should be rational
- More attention needs to be paid to quality control
- Periodic updating of knowledge is needed
- A mixed farming system is advantageous vis-à-vis monoculture say Agroforestry - Animal Husbandry or Aquaculture - Animal Husbandry.

Animal husbandry is an occupational culture of livestock production and lifeline of farming community. It is a mixture of breeding, feeding and management. These results are achieved only from 16 varieties of domesticated species out of over 3300 mammalians. In terms of value milk is the top most commodity produced in India followed by paddy, oil seed and wheat. Moreover, India is now the top milk producer in the world followed by USA. With 273.56 million bovines, India has about 25% of world stock. Of the bovine stock of India, only 18% are of good breed the rest being scrub. The task is very simple:

- Educate the farmers
- Sterilize unwanted stock
- Artificially inseminate to elevate breeds
- Resort to stall feeding. In dairy enterprise feed alone accounts for about 70% of total cost. The modern dairy cow is a product of many generations of genetic selection and is capable of tremendous milk production. The greater feed consumption and milk yield has increased the metabolic demands on the cow. Consequently, nutrients that traditionally were not supplemented must now be added to the diet.
- Where grazing is must, resort to rotational grazing
- Proper managerial practices. Newly born calves require more care. Cows in advanced stage of pregnancy should be isolated and provided additional palatable and laxative ration with clean water. Calving pen should be clean, well bedded, ventilated and disinfected. The weaned calf should be trained to drink milk from clean and disinfected pail. The milk from diseased animals should be disposed off properly so that it does not infect the flooring of milch animals.

Cow is a political animal in India. This should make the task easier in mobilizing resources. Unfortunately, so far there has been gross apathy in dealing with the problem of scrub cattle. Feral cattle also pose menace. Foresters have a pet whipping boy by way of overgrazing for failed plantations degraded forests. There is a surfeit of stringent laws to deal with the problem. Lacunae has been in implementation. Even in big cities cattle and pig rearing is rampant in posh residential colonies. With all the trappings of authority and power, it is a common sight to see cattle and pigs rummaging through muck in residential areas. Diseases like encephalitis has been attributed to such pigs. This again leads us to the sentiments expressed in Chapter 1. Indians as a collective entity are very fond of beating about the bush and not catching the bird for whatever reasons. If unwanted cattle are sterilized, over a course of time the population will be within manageable limits. It is unregulated breeding that is perpetuating the problem.

There is no dearth of knowledge base on veterinary science and pasture land development. IVRI at Izzatnagar and IGFRI Jhansi have done good research and plenty of literature is available. Goats are not inimical to preservation of greenery as popularly projected. If management is proper, goats and forests can coexist as demonstrated by research conducted at Goat Research Institute, Mathura and Agroforestry Research Institute, Jhansi. Sheep and goat are essentially dependent on rangelands, underdeveloped pastures and to some extent cultivated fodder. Tree leaves can sustain sheep and goats during adversity. About 60% of the total food requirement come from top feed. Ailanthus excelsa, Acacia nilotica, Prosopis cineraria, Azadirachta indica, Albizia lebbek, Acacia catechu, Hardwickia binata, Melia azadirach, Dalbergia sisoo, Grewia optiva, Ficus bengalensis, F. religiosa, Bauhinia variegate, tamarindus indica, Acacia leucophloea, and pods of Prosopis juliflora are relished. Shrubs like Capparis horrida, Zyzyphus mauritiana, Morus alba and Dichrostachys nutans also provide fodder leaves.

Forage crops are important as these provide major nutritional base in livestock feed. Shorghum is one of the major forage crops of India apart from being the most important non-leguminous kharif fodder crop. Single cut varieties of Shorghum are UP Chari 2, Pant Chari 3, MP Chari, HC 136 etc. which yield between 300 quintals/

ha.to 448 quitals/ha. Multicut varieties are SSG 59-3, Hara Sona, Safed Moti etc. which yield between 407 quintals/ha to 750 quintals/ha in green stage if all the prescriptions are followed. Lucerne is a leguminous fodder crop grown successfully throughout the country. It is annual as well as perennial, persistent, productive, heat and drought tolerant and provides fodder during summers when there is a shortage. The primary objectives of Lucerne cultivation should be to harvest plant at a stage that will provide satisfactory yield and still give the plant sufficient time to store adequate food reserves to maintain vigour for regrowth. Para grass (Bracharia mutica) is a nutritious high-yielding grass of the wet tropics where the climate remains warm and humid for the greater part of the year. Although the grass is water loving, it can withstand drought and other adverse conditions. Perennial Rye (Lolium parene) is a grass of temperate climate. It is an exotic and is being grown in India in Sikkim, J&K, HP and Uttaranchal. In subtropical region it can be grown for winter forage under irrigated conditions. It can be intercropped with subtropical grass Setaria kazungula.

As far as natural grasslands are concerned, it is estimated that 74% of earth is covered with water or ice, 9% is desert, 11% is forested and 6% is grassland. There are 255 genera and 1225 species of grasses in India. Foresters in India have been so much obsessed with trees that in the past attempts have been made to plant up natural grasslands with trees. The past cannot be undone, but in the interest of conserving biodiversity, natural grasslands should be left undisturbed because these are natural habitat of important fauna. But in production forests and agroforestry areas, grasses can constitute the ground flora for which many models are available with Indian Grassland and Fodder Research Institute, Jhansi. As per guesstimate, half of the livestock population of 270 million graze in forests besides extraction of green fodder from the forests to the tune of 175 to 200 million tonnes annually. To induce stall feeding behavioral changes are imperative.

With proper management goats are not a menace to vegetation

12 No Need To Grope In Dark

The biosphere is driven by the flux of renewable energies in sunlight, tidal momentum and deep earth heat. Human society draws energy directly from the environment, from short term storages like wood, soil and ground water and from long term storages of fossil fuels and minerals. In most systems, a significant portion of inflowing energy is degraded with smaller amounts transformed into higher quality energy.The overall scenario in India is very paradoxical. We boast about high availability of technical manpower. We have a credible space programme and plans to send astronauts to moon. On the other hand, electricity supply position can only be termed scandalous. For reasons we have to revert to Chapter 1. How to rectify the situation? First of all we have to devise a reasonably priced small solar electricity generating system with maintenance free battery. What is reasonably priced? It is a well-established principle of cost benefit analysis that social and economic penalties of alternative courses of action have also to be taken into consideration. The existing system of mega-sized power plants with long supply grids, bloated and corrupt infrastructure has failed in its objectives. People in urban areas are so habituated to fans, coolers, AC, TV, refrigerator etc. that during power cuts breakdown of law and order are frequent. Productivity goes down. Commercial activity suffers. There is a flourishing business in diesel generators and inverters. Diesel fumes cause environmental pollution with consequent health hazards. Most of the power corporations are incurring losses. Long transmission lines sometimes require forest clearance. There are thefts of aluminum wire rampant with additional burden on police and courts. Due to high tariffs and hassles related to legal connections, power thefts are common. Thus, the moral fabric of our society is being undermined with incalculable adverse results. Will it not be cost effective to import sturdy solar power generating systems from Israel, California, Texas or from wherever if locally manufactured ones are not performing well. Are we not importing T-92 tanks from Russia when indigenous Arjun failed. The list can be expanded with Agni missile, jet fighters and so on.

Once we agree on the necessity of sturdy, compact and reliable solar electric generating system for every rural household, all other problems automatically get sorted out. If direct to home TV (DTH) transmission is legalized, rural population can be moulded through edutainment. Literacy rate will automatically be 100%. The prevalent notion of considering those who can sign their names as literates has to be discarded. Only those are literate who know where to find information and who can make rational decisions on the basis of available information. With edutainment available at will, surplus time can be fruitfully utilized. With knowledge, the standard of life will go up. Subsidy is now a dirty word. Financing may be a better option. After all, even non-essential luxury item like car is now available to urban elites on 100% finance with repayment periods as long as 7 years – all to benefit some multinationals. Renewable sources of energy has been the stepchild of Indian planning system. Renewables form just 3% of the total installed power generation capacity in the country. There are about 80,000 unelectrified villages in India out of which 18,000 cannot be connected by grid. Renewable energy has to be the source of power for the rural marginalized. Companies like SELCO have done good work in parts of Karnataka in collaboration with some banks by penetrating into unelectrified villages without the benefit of subsidy. Even low income people are willing to pay a high amount as long as they receive quality power. It is a misconception that poor cannot pay for energy. India receives 5000 trillion kilowatt hours per year of solar energy, estimated at 20 mw per sq. km. The capital cost of installing a PV system is Rs 5000/- per watt. As a result, the life cycle cost of PV generated electricity is Rs10/- per kilowatt hour. If the costs involved in alternatives is kept in consideration, PV generated electricity will be worthwhile. Now low cost PV generators are in market which can charge the maintenance free batteries of small gadgets like mobiles, TV, DTH etc. So, villagers need not be dependant on erratic power supply for these gadgets which contribute a lot in improving quality of life and upliftment of society.

Cooked food is a basic necessity. A miniscule section of population has access to cooking gas. Though a user friendly item, transportation and handling costs are high. A lot of subsidy is involved which helps only urban elites. Solar cookers are much better alternatives. It did not catch up due to certain constraints which can be easily overcome. On cloudy days and during night the proposition is not feasible. One solution is to use solar cookers with electric back up. Coupled with solar generators described above, one can have wholesome food throughout the year and round the clock. Another solution is to use the solar cooker during sunshine hours for boiling, roasting, baking and drying purposes and to store the semi-cooked food in food grade polyurathene bags hermetically sealed with a tongue and groove arrangement and lined with bioceramic crystals which has a property of emitting far infrared rays (FIR). FIR rays do not permit bacteria, virus and fungus to proliferate. Thus the stored semi-cooked food does not become stale for a few days without even the benefit of refrigeration. Just before consumption, the semi-cooked food can be fried as per taste on ceramic sticks sprinkled with few drops of propane. These contraptions are lightweight and economical and are being used by trekkers and campers. Till such time, such contraptions are unavailable, conventional stoves, small cylinders of LPG or electric heaters can be used for giving final touches to the dish. Lightweight models of solar cookers are available in metallic and fibre reinforced plastic (FRP) bodies. Some of the advantages of solar cookers are:

- No fear of overheating or burning of food
- No environmental pollution
- Saves fuel and time
- Acts as a hot case
- Useful for boiling, roasting, baking and drying purposes
- Recovers cost in about 2 years
- Food retains nutrients due to slow cooking
- Used as a supplementary cooking device it can save kerosene, LPG or firewood. Those nations or organizations worried about carbon sequestration can put in the seed money for loans to purchase solar cookers.

The potential for biomass gassifiers in India is 19,500 mw. As against this, the installed capacity is just 400 mw. Biomass gasification is one of the most promising energy technologies for rural application. Most gassifiers use gas and diesel in the ratio of 80:20. One of the success stories of biomass gasification is in Taragram in Orcha in Madhya Pradesh of Development Alternatives. This 100 kw plant was designed by the Indian Institute of Science, Bangalore. Cost was only Rs 22,00,000/- and is now four years old. It requires one ton of biomass per day. Projects fail because no study is done to understand the biomass availability and possibility of growing plantations.

The potential of wind energy in India is 45,000 mw whereas the installed capacity is 1500 mw. The siting of wind farms is crucial and some outmoded restrictions on reactive power need to be removed.

In spite of the high fangled ideas enunciated in foregoing paras, kerosene remains the staple source of light in rural areas and will remains so in foreseeable future. The traditional hurricane lantern is the preferred contraption. Guesstimate is that 100 million such lanterns are in use in India. Noorie lantern designed by NARI (Nimbkar Agricultural Research Institute is an innovative device which can replace the traditional hurricane lantern. It is a pressurized mantle lantern which produces light output of 1250 to 1300 lumen (equivalent to output from a 100 watt bulb). This multi-fuel lamp can run on kerosene, diesel as well as ethyl alcohol producing similar light output. It is superior to existing pressurized kerosene lanterns (Petromax) available in the market. It produces light output equivalent to that from the petromax but consumes 60% of the kerosene that petromax requires. Noorie doubles up as a cooking device. This should be linked with what has been said for solar cooker i.e. roasted, boiled or semi-cooked food from solar cooker can be fried on Noorie. Technology Informatics Design Endeavour (TIDE) has installed energy efficient community cooking devices at some places.

Where does all this discourse leave the headloaders? Actually they are an exploited lot. If the expenditure in terms of human energy is compared with the output from that bundle of firewood, the effort is far from cost effective. There is a whole series of middlemen and officials who take their cut. A miniscule part of the firewood so extracted is used for cooking the food in the hearth of the headloader. The balance is collected in the premises of some middleman to be sold to brick kilns, tea shops and for consumption in towns. The headloader is in the transaction because he has no alternative avenue for gainful employment. Meagre earning from firewood collection keep his body and soul together. Once he or she is enlightened about more gainful and less energy intensive avenues of occupation, the incentive for scouring the forest floor for firewood will vanish. Here comes the role of education.

13 Wanted Spin Doctors – Anybody Willing To Pick Up The Gauntlets

Bharat Ratna President APJ Abdul Kalam speaks of making India a developed country by 2020. He has a vision document. For accomplishing this vision, the system has to be sanitized first. Let us face the facts. Of the 1030 plus million population, about 80% are leading a zombie-like existence. They are passing through their life cycles as if in trance, being led by some smart alecs – some being downright crooks, others so fond of trappings of power, that they are not averse to collaborating with the former. About 1% of population is highly motivated and productive which works out to a sizeable number - bigger than total population of many developed countries. The rest of population is riding piggy back on the 81% described earlier. The numerical strength of NRIs is more than the combined populations of several developed nations. Freed from the stifling and debilitating milieu of their native land, the NRIs have individually and collectively performed very well. Affluent NRIs are contributing significantly to the inflow of foreign exchange. Nature is bountiful and the zombies manage to produce enough to give India the trappings of a regional super power. What about human development index? The Green India report brought out by Tata Energy Research Institute (TERI) paints an appalling picture. The availability of fresh water had gone down from 6000 million cubic metres to 2300 million cubic metres. The cumulative loss from deforestation, erosion, water-logging and salinity had risen from Rs 8300 crores to Rs 23200 crores. (These figures are for the period 1947 to 1997). Two and a half million people had died prematurely in 1997 from adverse effects of air pollution. We are tom tomming about so called increase in forest cover from 63.34 million hectares in 1997 to 63.73 million hectares in 1999. These figures are illusory because forest cover is assessed from satellite imagery. The catch is that technology has improved. The resolution of satellite camera is now 5m X 5m against 25m X 25m previously. So, smaller patches of greenery are also registered which were previously recorded as blank. Already we have seen dangerously contaminated water making way to our dining tables. Even deadlier pesticides have now entered our food chain. Only 52% of households have pucca roofs over their heads, 60% have no drinking water in their premises and 122 million lack basic sanitation. Do we want to be a developed nation just replicating the present day scenario? Out of the economically active population of 550 million at present, only 27.53 million have benefit of employment in organized sector. 249.46 million are either employed or underemployed in the unorganized sector. Services of the balance 272.71 million remain unutilized. Over 29% of population continues to suffer under the clutches of poverty and hunger. All the economic activities underway cannot produce dramatic results to pull the 300 million out of the morass of utter destitution. A study "District Level Deprivation in The New Millennium" conducted jointly by Rajeev Gandhi Institute of Corporate Studies and Indices Analytics paints a similar picture. The criteria adopted for classification are as under – same as adopted by UNDP.

1. Poverty
2. Hunger
3. Literacy
4. Vaccination
5. Child Mortality
6. Basic Education

As per this study, 69 districts were below the benchmark in at least four out of above six criteria whereas 76 other districts were found deficient in standards set by UNDP. Some of the findings are listed in table below:

Bihar	Jharkhand	Orissa	UP	MP	Arunachal Pradesh	Karnataka
Twentysix out of 37 districts are in the list described above. In 36 districts there is no facility for vaccination. In 31 districts, facility for admission in primary schools is negligible. 30 districts are at the bottom of the rung so far literacy is concerned.	Ten districts out of 22 are in the list. Fourteen are affected by starvation, 12 by abject poverty whereas in 30 there is little or no facility for vaccination. Ten districts have been listed as below standards set.	In 23 districts out of 30, there is abject poverty and starvation. In 30 districts there is little or no facility for vaccination. Two districts have been listed as worst whereas 10 are below standard set.	In 54 districts the ratio of admission in schools is below benchmark, in 27 there is illiteracy, in 36 there is little facility of vaccination and in 18 there is abject poverty.	Six districts are in the list. There is abject poverty in 22 and in 39 , rate of infant mortality is abnormally high.	Nine districts are affected by starvation, in 8 there is little facility for vaccination and 7 districts are below benchmark set for literacy.	In this silicon valley state, Gulbarga district is in the category of worst. Every second man is below poverty line.

Some salient features of this study indicate that in India there are:

1. 268.1 million below poverty line which is equal to the population of USA or Indonesia.
2. 27.3 million are suffering from pangs of hunger or starvation which is about equal to the population of Venezuela or Morocco. In other words 3 out of every 100 persons remain hungry.
3. 302.5 million are illiterate which is about equal to the population of Russia. This is due to poverty as well as root cause of poverty. Half the illiterates of the world are in India.
4. 1.8 million cases of child mortality are estimated annually in India. This can be compared to the population of Panama or Kuwait.
5. 12.2 million children have no access to vaccination facilities. This figure can be compared with the population of Angola or Cambodia.

Most of the backward districts have poor road network. India may boast of medical tourism but 6 out of 10 children are born without medical aid. No wonder that many of the backward districts are affected with Naxalites. Most of the development funds have been sucked in the blotting pad of maladministration.

Japan and Israel have been successful in providing full employment to all their economically active population. The per capita availability of land resources in India is double that of Israel. The per capita availability of water in India is nearly six times that of Israel. If one drives from Tel Aviv to the Egyptian border, about halfway the land changes to desert, but the farming does not stop. The land is covered with hundreds of hectares of climate controlled computerized automatically irrigated greenhouses growing tomatoes,

A green house in Israel

A computerized experimental green house in Israel

eggplants, flowers, decorative house plants and so on. Even further south near Eilat, farmers grow melons, watermelons, cauliflowers and all sorts of fruits and vegetables in a desert that receives rainfall of less than 50 mm. In Kibbutz settlements, large herds of dairy cattle can be seen where cows have not seen a blade of grass. Fish farms are flourishing in desert that produce sea fish. The Israeli's use the latest remote sensing technologies including water tensionmeters for testing soil humidity as well as carbon dioxide meters and other such gadgets which are computer controlled. Basically, once the field or greenhouse is programmed, it will receive the appropriate amount of water, fertilizer, CO2, air humidity and even right wavelengths of light in order to maximize output. In1950, in Israel one agricultural worker was able to feed 17 people. The figure for 1994 was 90 people. Solar energy is available in abundance in India. Only if we can harness these resources to the optimum, it will be the biggest poverty alleviation project ever undertaken. As per the National Forest Policy of India 1988, 33% of geographical area should be under forest/tree cover. Though the exact percentage can only be worked out after the exercise suggested below, but as per guesstimate, this figure of 33% may be already in place. Forest does not only mean areas covered with trees. Blank, rocky, waterlogged, snowcovered and sandy areas are important ecosystems and constitute forests if legally so defined. Due to traditional mindset, foresters endeavour to plant trees in all such habitats. This tendency has to be curbed. Then, individual trees standing in linear strips and private fields have crowns which cast footprints on the ground below. If all such footprints are added together summed up with legally defined forest area, then the figure may well neigh touch the magical figure of 33%. The task is not too difficult. There are commercial US satellites in orbit which are recording aerial photographs with resolution of 1mX1m. That means all individual trees standing anywhere in India, can be identified and their crown cover measured. Still, we should go beyond this sterile debate and try for optimum production per unit area per unit time.

At this juncture it will be worthwhile to juxtapose in Table I some statistics relating to India and Israel.

Table I

Sl.No.	Item	Israel	India
1.	Land Area in Sq.Km.	20,770	3.29 million
2.	Population	48,22.000	1030 million
3.	Labour force	15,53,000	200 million
4.	GNP per capita	US$13,000	US$500

What characterizes Israeli agriculture, making it one of the most advanced and intensive in the world is a combination of several factors: young in years, lack of tradition, the motivation of the first pioneers who came to Palestine some 100 years ago to change the mode of life of the Jewish people in the diaspora and no less important – the limited natural resources which encouraged the development of modern sophisticated and quite often even unique agriculture technologies, particularly in the utilization of water and soil resources. The first settlers who emigrated to Palestine were highly motivated ideologists charged with a spirit to do things differently and were mostly from high socio-economic background. It was realized that the agricultural structure should be diversified in order to have a surplus both in social and economic aspects. The linking gap between farmer and scientist is very narrow and the farmers are willing to be guinea pigs. Farmers comprise only 3.6% of the country's population and export of agricultural produce is to the tune of US$1.2 billion. Israel is a leading nation in the use of solar energy.

The so called waste land, or wasted land, in India is estimated to be 175.06 million hectares. This is 84 times the total geographical area of Israel. If the productivity per unit area per unit time from this wasted land in India is brought at par with that of Israel, then the export potential can cross US$168 billion at a conservative estimate. For this to happen, the level of motivation has to be substantially enhanced and mindset has to be changed. It is fortunate as well as unfortunate that foresters in India are stuck up with the appendage of Consevator – from Assistant to Principal Chief with Deputy, plain and simple Conservator and Chief Conservators in between. It is good that foresters are engaged in the task of conserving our heritage. But the unfortunate part is that in undertaking this noble task, foresters are very conservative in their approach. They are not very open minded about emerging trends and technologies. The blinkered vision needs to be shed. What is needed is a kaleidoscopic vision. Functioning of human mind and parachute is akin. Both work best when open. Choice is ours.

In a supposedly corrupt nation like India, the three companies having the highest market cap – Wipro, Hindustan Lever and Infosys – are companies with the highest perceived integrity. **Leadership = Character X Competence**. Leaders are known:

- Less for what they say and more for what they deliver
- Less by their title and position and more by their expertise and competence
- Less by what they control and more by what they shape
- Less by the goals they set and more by mindsets they build
- Both for personal character and for exceptional organizational competencies

Trustworthy leaders have the personal habits, values, traits and competencies to engender trust and commitment from those who take their direction. If we balance information and knowledge with wisdom, rights with duties, ends with means a new renaissance;

a new creative vision of better world awaits us. It will need widest possible mobilization of ideas, knowledge, skills coupled with ethics. Rather ethics is a prerequisite.

Some suggested ways of increasing gainful employment are:

- Increase export of medicinal herbs. As per WHO over 80% of world's population rely on traditional system of medicines, largely plant based, to meet primary health care needs. Rs 550 crore worth of medicinal herbs were exported from India during 1977. This is only 0.5% of the global export market. If the export figures can be increased to Rs 5000 crores annually, 20 million families can be gainfully employed.
- Value added dairy products can provide gainful employment to 80 million. (The rider is we shun the criminal practice of fobbing off synthetic milk to gullible consumers while responsible authorities look the other way).
- Prevention of waste. Nearly 30% fruit and vegetable production is wasted. Hermetically sealed food grade polyurathene bags can help in the initial stages of harvesting. Subsequently, these can be canned and preserved. Hygienically packaged natural drinks like coconut water and fruit juices can substitute overpriced and often unhealthy soft drinks flooding the market.
- Job-oriented education. This need not be only formal. Edutainment should be resorted to. Customer as a co-worker is also emerging as a popular concept.
- Ecotourism. This will require shunning of unethical practices and improvement in unsanitary conditions. Availability of clean drinking water will go a long way in encouraging tourists to visit off-beat spots.If taken up as a commercial venture, the constraints can be overcome. Where ever accommodation is deficient, use of caravans will be advisable. Approach roads

can be improved with private sector funding to be recouped through toll barriers.

- There are umpteen number of alternative avenues of more gainful employment which can be adopted by rural poor – specially women. Rearing of ornamental fish is one the global market for which is five billion dollars. Share of India is only 0.007%. In a project in South 24 Pargana district of West Bengal, one fisherwoman is earning Rs 5000/- per month since she shifted to rearing ornamental fish which is three times what she was earning previously. Medicinal and edible mushrooms with buy-back guarantee are other lucrative items.
- Detractors will lament about source of financing. At least one bank – Vijaya Bank – has some financing propositions for high-tech agriculture. The bank defines Hi – Tech agriculture as adoption of new and innovative technology on commercial scale for increasing production, productivity, quality and value addition to agricultural produce. Finance is available for floriculture, mushroom cultivation, tissue culture, cultivation of medicinal and aromatic plants, strawberry and gherkin, intensive aquaculture, frog farming, vermiculture, hydroponics, culturing of Azotobacter, Rhizobium, Mycorrhiza, blue green algae, Spirulina etc. for manufacture of bio fertilizer/use as food/pharmaceutical industry, culturing of Bacillus thuringensis etc. for use as bio pesticides, environment controlled poultry, hatcheries and breeding farms etc. Agri finance branches are manned by technically qualified and experienced officers and deal exclusively with agricultural proposals guiding entrepreneurs regarding project preparation, sources and availability of subsidy, policy guidelines of government, availability of forward and backward linkages and act as preliminary consultants. All kinds of banking services are offered by these branches exclusively to the client/prospective

customers so that entrepreneurs get the best and quickest personalized service including counseling.

Talking of tourism, our cities are first to be exposed to the gaze of foreign and interstate tourists. No doubt, greenery is soothing to eyes. But, in urban context aesthetics is equally, if not more important. Foresters and arboriculturists in India have done next to nothing about landscape architecture in urban context. Funds are wangled in the name of greening of cities, pits dug in pavements, road dividers and traffic islands and planting carried out without giving any thought to eventual height, extent of crown cover, foliage and flowers. Often species with a tendency to grow tall are planted underneath electric transmission lines. After a few years personnel of electric supply company come and hack and lop the trees indiscriminately. Net results are eyesores. By this time forest department has washed its hands about old plantations. Hunting for new targets is a more lucrative proposition for them. In advanced countries, tree architecture is an important feature of urban forestry. Avenue trees are regularly pruned to desired shape. Priorities are also lopsided. Hundreds of years old trees are felled without remorse for widening of roads. As an eyewash, shrubby or quick growing trees are planted on already shrunken footpaths which are subject to encroachment by hawkers. Result is failure or stunted bushes. After, a few years people loose sight of the historical perspective and second round of efforts gather momentum for release of fresh funds. Truly, government projects about macro investments with micro results. There is lot of hullabaloo about green islands and breathing lungs of cities. One example will illustrate the paradoxical situation between what we say and what we do. In Lucknow one area was to be developed for holding large political gatherings. One patch was selected within city limits. More than 5000 trees – some more than 150 years old — of different species were standing in the tract. All these trees were ruthlessly felled or bulldozed. Star TV channel telecast the event. A person at the helm of affairs, ran away from his office room waving hands and shouting no comments when confronted by TV camera crew. Wheelers and dealers in administration got active and the news was blacked out from media. Some public spirited persons are still agitating the issue of this Ambedkar Ralley Sthal at appropriate forums. Some heads may or may not roll. The fact remains that the green cover has been lost for ever and newly planted saplings can never compensate the old uprooted trees in number and crown cover. Contrast this to the situation that emerged in Athens. Four hundred olive trees were required to be felled for 2004 Olympics. Green VSOs (Voluntary Social Organisation) were up in arms. Ultimately, the compromise solution was that the uprooted trees were to be safely preserved and replanted after Olympics.

If the drift is to be arrested, persons have to come forward to pick up politically orphaned ideas some of which have been narrated in foregoing chapters. Instead of cursing the darkness, it is better to light a candle. Our time on earth is short. We are only trustees of what we think we have. Nothing is really ours. But if we use our abilities to help other people, then we can make a difference. What we do for others is the only thing we take with us after we leave our mortal body. Actions speak louder than words. Some VSOs and industrial houses are doing yeoman's service. On the other hand there are lots of black sheep amongst VSOs and NGOs. The country is vast and efforts of good NGOs and VSOs are a drop in the ocean. All like-minded persons should take a pledge not to countenance social evils and to harness their resources to set matters right. With tools like human behavioral sciences, digital multimedia, internet, telecommunication and above all lateral thinking, things are bound to take a turn for the better. Lateral thinking needs to be elaborated. Our education system is based on the ideas developed by Plato, Socrates and Aristotle in which stress is laid on linear thinking i.e. one is trained to proceed from point one to two and so on till the goal say point eight is reached. If any insurmountable obstacle is encountered at say point three, one is stumped and the tendency is to wait till things get sorted out. During recent times, as opposed to linear thinking, Edward de Bono has advocated lateral thinking i.e. one should explore ways and means of reaching point eight which is the goal by bypassing points four, five, six etc. There is nothing new in this thinking. Many top-class military geniuses, captains of industry and explorers have been devotees of lateral thinking without the benefit of theoretical systematizing. Many countries like Israel have been able to overcome seemingly insurmountable problems to come up on top.

Perorations

1. The endeavour in Indian context should be to turn the zombies into rational beings. In a democracy majority decides the course of action. The system does not take into account whether the zombies are puppets on strings being manipulated by unscrupulous elements. The result is that we have the dubious distinction of being at the receiving end of grand gift of poor governance. Perhaps the adage that a society gets the government it deserves is after all proving to be self fulfilling. Sands of time are slipping relentlessly. **Cognoscenti have to act fast before nemesis catches up.**

2. Not statistics: vision is needed. Agri diversification, innovative technologies and new market linkages could generate jobs.

3. Shed tunneled vision.

4. With the fast changing world scenario, India must be prepared to accept the challenge in taking a leading role in the field of natural products with emphasis on value addition.

5. Resort to super intensive agriculture (including forestry) which implies making the absolute maximum of every element of land, water, equipment, know-how and manpower and combining them to achieve more than the sum of parts.

6. If one is facing steamroller technologies narrated in foregoing discourse, one is either part of the steamroller or becomes part of the road.

7. Above all, induce Indians to be socially responsible citizens. An organization and society is an open, dynamic, multigoal seeking, purposeful system that has elements of concreteness and abstraction. They not only are affected by the pervading environment but also affect it.

And The (R)evolution Continues.........

Index

A

Acacia nilotica 43
Acacia Senegal 43
Agrisilviculture 66
Alachlor 31
Atrazin 31
Azadirachta indica 43

B

Babool 43
Ber 43

C

C. garaf 43
Caligonum polygonoides 43
Capparis deciduas 43
Cordia myxa 43
Cryopreservation 52

D

Diuron 31

E

Echnocloa colona 45
Electric Trimmer 35
Exotics 41

F

Forest Fires 78

G

Genetic modification 56
Glyphosate 31
Goonda 43
Goondi 43

H

Holy Cow 93

I

Imli 45

J

Jungal jalebi 43

K

Ker 43
Khejri 43
Kumat 43

L

Leucaena leucocephala 45

M

macroproliferation 38
Metribuzin 31
mini sprinklers 17
Moringa oleifera 43
Multi Tier Planting 65

N

Natives 41
Neem 43

P

Paraquat 31

Peelu 43
Phog 43
Pithecellobium dulce 43
Portable Solar Electricity Generator 104
Prometheus Bound 1
Prosopis sineraria 43

Q

Quadrupeds 93

R

Rain gun 16

S

S. persica 43
Sahjan 43
Salvadora oleoides 43
Soma 45
Spin Doctors 105

T

Tamarindus indica 45
Terminalia arjuna 46
Tractor mounted sprayers 17

U

Ultra Sonic Mist Generator 15

V

Vetiver 74

Z

Zizyphus nummularia 43